Actuaries' Su

D0380852

How to Succeed in One
of the Most Desirable
Professions

Actuaries' Survival Guide

How to Succeed in One of the Most Desirable Professions

Fred E. Szabo

Department of Mathematics and Statistics
Concordia University

ELSEVIER
ACADEMIC
PRESS

AMSTERDAM • BOSTON • HEIDELBERG • LONDON
NEW YORK • OXFORD • PARIS • SAN DIEGO
SAN FRANCISCO • SINGAPORE • SYDNEY • TOKYO

Wayne Chen of MBC Design in Montreal created the cover of the book. It is a compass over a world map, symbolizing the book's role as a guide through the complex world of actuaries. The colors are chosen to represent a sense of adventure and discovery associated with the reading of the book and with the embarking on an exciting and challenging career.

Elsevier Academic Press
525 B Street, Suite 1900, San Diego, California 92101-4495, USA
84 Theobald's Road, London WC1X 8RR, UK

This book is printed on acid-free paper. ⊗

Library of Congress Cataloging-in-Publication Data

Szabo, Fred.
  Actuarial survival guide : how to succeed in one of the most desirable professions / Fred E. Szabo.
     p. cm.
  Includes index.
  ISBN 0-12-680146-0
  1. Actuaries–Vocational guidance. I. Title.
HG8781.S93 2004
368'.01'023–dc22

                                                                    2004000029

British Library Cataloguing in Publication Data
A catalogue record for this book is available from the British Library

ISBN: 0-12-680146-0

For all information on all Academic Press publications
visit our Web site at www.academicpress.com

Printed in the United States of America
03   04   05   06   07   08   9   8   7   6   5   4   3   2   1

Prediction is very difficult, especially about the future.
Niels Bohr, Physicist and Nobel Laureate

CONTENTS

ACKNOWLEDGMENTS

I would like to thank the following actuaries, actuarial students, mathematicians, economists, consultants, and career experts for having participated in the design and completion of the actuarial survey on which the hands-on material in the book is based:

Jonathan Bilbul (ING Canada), Marie-Andrée Boucher (Ernst&Young), David Campbell (Manulife Financial), Steve Cohen (ING Canada), François Dauphin (New England Financial), Karine Desruisseaux (Mercer Human Resource Consulting), Pierre Dionne (CCR Canada), Norman Dreger (Mercer Human Resource Consulting), Jean Drouin (National Bank of Canada), Julie Duchèsne (Mercer Human Resource Consulting), Louis Durocher (IAO Actuarial Consulting), Amélie Girard (Towers Perrin), Philippe Gosselin (ING Canada), Karine Julien (Actuarial Student), David Laskey (Hannover Re), Dany Lemay (Towers Perrin), Erik Levy (Bain & Company), Jean-Grégoire Morand (Mercer Investment Consulting), Paul Morrison (GGY Inc.), Lambert Morvan (Fairfax Financial Holdings), Dylan Moser (Actuarial Student), Céline Ng Tong (Actuarial Student), Marc Parisien (GGY Inc.), Karlene Parker (Hartford Life), Caroline Piché (Mercer Human Resource Consulting), Étienne Plante-Dubé (ING), Elisabeth Prince (Ernst & Young), Graham Rogers (London Life), Martin Rondeau (Mercer Human Resource Consulting), Siobhain Sisk (Mercer Human Resource Consulting), Mariane Takahashi (Actuarial Student), Véronique Tanguay (Towers Perrin Asset Consulting Services), Chantale Taylor (Consulting Services), and Ghislaine Yelle (Career Coach and Human Resources Consultant). The information that they have supplied in the survey or by direct communication, and that supplied by others, is reproduced in this book in anonymous and sometimes paraphrased form because the survey was designed to guarantee the confidentiality of the answers.

The survey was completed on the understanding that the opinions expressed are personal and should not be construed as representing the views of the companies where many of the respondents are employed.

I would like to acknowledge, in particular, the contributions of the reviewers of this project. They include Dale Borowiak (University of Akron), Bruce Edwards (University of Florida), Louis Friedler (Arcadia University), Jos Garrido (Concordia University), Brian Hearsey (Lebanon Valley College), Jon Kane (University of West Washignton), Stuart Klugman (Drake University), Jean Lemaire (University of Pennsylvania), Murray Lieb (New Jersey Institute of Technology), Vania Mascioni (Western Washington University), Charles Moore (Kansas State University), Kent Morrison (California Polytechnic State University), Walter Peigorsch (University of South Carolina), Gabor Szekeley (Bowling Green State University), Charles Vinsonhaler (University of Hawaii), and Bostwick Wyman (Ohio State University), as well several anonymous referees who provided guidance with the design of the project. I would like to thank all of them for their constructive comments. They will recognize traces of their ideas throughout the text.

The Society of Actuaries has granted me permission to build chapter 2 around sample questions and answers from the May 2001 examinations in Courses 1 through 4, the Casualty Actuarial Society has granted me permission to include the results of their survey on CAS professional skills, and the International Actuarial Association has permitted me to include its list of competency areas of actuaries. I hereby express my sincere thanks to them.

Others have provided direct information in other forms, and I am grateful to them. They include Michelle Aspery (Institute of Actuaries of Australia), Malcolm Campbell (COO Skandia Offshore Business), Maria da Luz Fialho (Portuguese Institute of Actuaries), Peter Diethelm (Association Suisse des Actuaires), Wim Els (Actuarial Society of South Africa), Yves Guérard (International Actuarial Association), Caroline Henderson-Brown (The Actuarial Profession), Betty-Joe Hill (Royal & SunAlliance), Curtis E. Huntington (University of Michigan), Liyaquat Khan (Actuarial Society of India), Pat Kum (Actuarial Society of Hong Kong), Dr. Eduardo Melinsky (University of Buenos Aires), Dr. Mario Perelman (Argentinian Institute of Actuaries), Dr. Jukka Rantala (University of Helsinki), Loredana Rocchi (Italian Institute of Actuaries), Deborah R. Rose (Faculty and Institute of Actuaries), Dr. Rafael Moreno Ruiz (Universidad de Málaga), Nicole Séguin (International Actuarial Association), Martha Sikaras (Society of Actuaries), Elizabeth Smith (Casualty Actuarial Society), Stuart Szabo (Global Corporate Finance, Deutsche Bank), Klaus Wegenkittl (Union Versicherungs-Aktiengesellschaft), Karin Wohlgemuth (Zurich Financial Services), Yew Khuen Yoon (Actuarial Society of Malaysia), Masaaki Yoshimura (Institute of Actuaries of Japan), and Aleshia Zionce (Society of Actuaries).

I would like to express my deep appreciation and gratitude to Dr. Harald Proppe, my colleague, and to Eric Hortop, my student, for having spent innumerable hours reading the manuscript and suggesting corrections and improvements.

A special thanks is reserved for Barbara Holland, my editor, who believed in the project and encouraged me to carry it out. Further thanks are due to Tom Singer (Academic Press) and the Production Team (Academic Press).

The anonymous survey was produced and evaluated with the help and guidance of Maggie Lattuca of the Concordia University Instructional and Information Technology Services Department using *Respondus* and *WebCT*.

This book was written in LaTeX using Scientific WorkPlace. I would like to thank Barry MacKichan and his team for continuing to produce and improve this unique scientific writing tool. The camera-ready copy was prepared with much love and care by Kolam Information Services, and I would like to thank Christine Brandt and her staff for their meticulous work.

Fred E. Szabo
July 2003

PREFACE

You are reading this book because you are thinking about the future. What would you like to do with your life? What career would allow you to fulfill your dreams of success? If you like mathematics, your choices have just become simpler. Consider becoming an actuary.

In the pages that follow, I will explain to you what actuaries are, what they do, and where they do it. I will also whet your appetite by explaining some of the exciting combinations of ideas, techniques, and skills involved in the day-to-day work of actuaries.

One of the key features of this book are the answers provided by over 50 actuaries and actuarial students in an electronic survey about the actuarial profession, sent to one 150 experts. The submitted answers are included in the book, with minimal editing to preserve their flavor and the spontaneity of the replies.

Another useful feature of the book is the inclusion of sample questions and answers from joint Society of Actuaries (SOA) and Casualty Actuarial Society (CAS) examinations. They are presented in a *look-and-feel* format. By browsing through these sections, you will get an idea of what the questions *look* like and get a *feel* for what the answers should be. Although the form and content of these examinations will change over time, the ideas and techniques presented in the quoted examples will give you an idea of the mind set of professional actuaries. Chances are that neither the questions nor their answers will make sense to you at this time. But by perusing their content and looking at the form of their answers, you will get a sense of what lies ahead. Nevertheless, this book is not a study guide. In order to pass the SOA and CAS examinations, you must use some of the techniques and study tools discussed in the Chapter 2. You will find a list of appropriate references on the websites listed in Appendix D. In the case of the SOA Courses 5 through 8, and the CAS Courses 5 through 9, you will only find summaries of the course descriptions. They will give you an idea of the content of these courses. You will quickly realize that the courses are based on ideas and techniques related to actuarial work *experience*. If you have managed to pass the first four foundation courses, you will have learned from your colleagues what to expect in these advanced courses and how to prepare for their examinations.

A third aspect of this book that you will find useful is a list of typical employers. The list is incomplete because there are thousands of public and private companies, as well as government agencies employing actuaries. It is based on personal contacts and suggestions received from respondents to the survey. As presented, the list is meant as a starting point for your personal research into actuarial employment. The details provided about different companies differ from employer to employer. The intention is to highlight different aspects of employment rather than giving an encyclopedic description of employment at particular companies. You can easily complete the sketches by consulting the cited websites.

This is a hands-on book. For more than ten years, I have been associated with actuarial students as the director of an actuarial cooperative program at Concordia University. Before writing this book, I consulted over 100 of my former students and their employers about what kind of introduction to their profession they would have liked to have had when they were making their career choices. Their answers form the background to this book.

The World of Actuaries

As you explore the world of actuaries, you will come across several sources of information. Societies such as the Society of Actuaries, the Casualty Actuarial Society, the Canadian Institute of Actuaries (CIA), the Faculty of Actuaries (FA) of Scotland and the Institute of Actuaries (IA) of England, as well as similar organizations in the rest of the world should be at the top of your list of primary sources. In Appendix D, you will find links to relevant websites. In addition, the section *Actuaries Around the World* in Chapter 1 contains information on what it takes to be an actuary in different countries.

What is an actuary? A mathematician, statistician, economist, investment banker, legal expert, accountant, or business expert? I will show you that an actuarial career involves elements of all of these professions and more. I will try to open the rich mosaic of actuarial life for you. As such, I would like this book to be more than a career guide. I would like it to be your career companion on the road to professional success.

When discussing this project with a senior actuary, I was told that writing a book about actuaries is like trying to shoot at a moving target. The book needs to be updated as soon as it is written. This fact made the project an even greater challenge and more exciting. It made me realize that my job was to concentrate on the big picture, the ideas and scenarios that unite this changing and dynamic world and give it permanence. The book in your hands is the result.

Much of the material integrated in this book is in the public domain and is available in bits and pieces through a multiplicity of published sources. However, it is widely scattered and incoherent and, as such, appears disjointed and overwhelmingly complex. One of the objectives of this project was to analyze the available information and build a coherent picture. In assembling the material for this book, I asked former students and some of their employers for comments on my plan. Here is what they had to say:

Q **Which topics in this book do you consider to be the most important and why?**

Answer Non-actuarial opportunities for students who enter an actuarial program at university.

Answer Understanding the full range of career options enables a better career choice with increased odds of job satisfaction and high-profile success.

Answer Technical skills (more important than interpersonal skills), internships (the best way to learn about the profession and find a full-time job), career profiles (since there are more profiles than people can imagine), actuarial recruiting (not always well known).

Answer I would say, equally, employers and careers because we don't learn that in school. Sometimes, teachers have not worked in companies, so they may not be familiar with this information. It can be a concern at any level, from high school to university.

Answer The chapter about the different career possibilities. It think this is important because as long as we are not working in a particular field, it is really hard to have an idea of what it is about. A clear description of these career opportunities would be really helpful for choosing both an internship and a permanent job.

Answer The skills. A lot of people do not know if an actuarial career is for them. Other important topics are the profession and the industry. I believe that the actuarial profession is not for everyone and that one has to love it to be happy and successful in it. Knowing what an actuarial career is all about is important before embarking on it. Other important topics are the professional courses; becoming an actuary is a lot of work and being aware of the studying that it requires is important before taking the decision to opt for an actuarial career.

Answer Real world applications of exam material, just so that students taking the exams feel that what they are learning is actually useful.

Answer The differences between different actuarial fields. For example, risk management is an area of interest to me. Another topic of interest to me is non-traditional areas of work.

Answer You need to know what is required to be a *good* actuary and whether you are suited for this career. Moreover, the choice of SOA [pensions, health, finance, consulting, etc.] or CAS [property and casualty insurance, etc.] is extremely important. I was in the SOA stream when I thought of changing fields. Once I had the opportunity to work with a P/C [Property and casualty] insurer, I saw the light and decided to switch to P/C (and I love it). I must say that school did not help me in making this right decision. Not enough information was given to students about the CAS option.

Answer The part concerning the SOA and CAS courses since, in my opinion, they are the most important aspect of an actuarial education.

Answer The chapter dealing with SOA and CAS career choices and a list of leading employers interest me the most.

Answer An actuarial background is a great asset for many more jobs than people might think. I am sure that in the near future only a small percentage of students graduating in actuarial mathematics will actually do actuarial work. They might not be typical actuaries but can easily become, given the necessary personal skills, great leaders in different areas of the business world. An actuarial mathematics background opens the door to a vast world of opportunities in the marketplace.

Answer It would be interesting to have a listing of companies with the type of jobs they are offering.

Note: Throughout this book, we use the acronyms SOA and CAS both as names for the *Society of Actuaries* and the *Casualty Actuarial Society* and as designations for *careers* for which an Associateship or Fellowship in these or similar societies is normally required.

▶ Chapter 1

ACTUARIAL CAREERS

1.1   Professional Options

The word *actuary* comes from the Latin word *actuarius*, which referred to short-hand writers in the days when things like typewriters and computers hadn't even been thought of. Today, actuaries work for insurance companies, consulting firms, government departments, financial institutions, and other agencies. They provide crucial predictive data upon which major business decisions are based. True to their historical roots, actuaries still use a rather extensive shorthand for many of the special mathematical functions required for this work. (*See* Reference 5, Pages 687–691, and Reference 18, Pages 123–131, Appendix F.) The sample questions and answers for Courses 2 and 3 in Chapter 2 illustrate some of the currently used actuarial symbols listed in Appendix E. The symbols are an amazingly rich combination of right and left subscripts and superscripts, attached to designated upper- and lower-case Roman and Greek letters.

Actuarial science is an exciting, always-changing profession, based on fields such as mathematics, probability and statistics, economics, finance, law, and business. Most actuaries require knowledge and understanding of all of these fields and more. To ensure that this is really the case, all actuaries must pass special examinations before being recognized as members of the profession. To perform their duties effectively, actuaries must also keep abreast of economic and social trends, as well as being up-to-date on legislation governing areas such as finance, business, healthcare, and insurance.

No doubt you have heard about the actuarial examinations you need to pass to become an *Associate* or *Fellow* of one of the actuarial societies. Often full-time employees in actuarial firms who are still engaged in the examination-writing process are distinguished from Associates and Fellows by being referred to as

Students. The efforts required to succeed in these examinations are in many ways analogous to those required to become a doctor, lawyer, or other high-ranking professional. So are the rewards. For several years now, the Jobs Rated Almanac has considered an actuarial career to be one of the most desirable professions in America (*see* Reference 13).

Actuaries are experts in the assessment and management of risk. Traditionally, the risks managed by them have been insurance and pension funding risks, although the management of business risks is also among the responsibilities of insurance actuaries. So is the insurance of insurance, known as *reinsurance*. Moreover, many actuaries are now also managing asset-related risks in merchant banks and consulting firms. This bodes well for the long-term future of the profession, since risks of all kinds will always be with us. However, as you will see later on in this book, the day-to-day activities of an actuary depend very much on the sector of the financial services industry where the actuary works.

Actuaries are often chosen to be general managers in insurance companies. This is because upper management and boards of directors have a high regard for the knowledge and skills of actuaries, and because the need of a company to maintain its financial integrity makes an actuary's numerical skills invaluable.

Actuarial Terms, Acronyms, and Definitions

As you read on, you will quickly discover that actuarial science is full of technical terms, acronyms, and definitions. This book is not the place for explaining them in detail, because the definitions involved are readily available in textbooks and on the Internet. The main objective of this book is to introduce you to the career opportunities that exist in the actuarial world and to sketch for you the steps required to enter that world. For this reason, most of the technical material in the book is provided only in illustrative and summary form. Consider it a detailed roadmap to the relevant topics in mathematics, business, and statistics. It is merely meant to help you identify the range of knowledge involved in actuarial work. The study of the mentioned topics requires specialized sources and tools. The reference section at the end of the book provides you with the necessary pointers.

Actuaries can be grouped in different ways. As their functions change in response to changes in the world around us, the distinctions become less sharp. However, the following categories of employment will give you an initial idea.

Valuation Actuaries

Reserves are important to the long-term financial health of a company. Because insurance companies are dealing with events that are uncertain in time and amount, they must put aside what they consider to be the most likely amount of money they will need to pay future claims and expenses, and then put aside a little more, just in case. The role of *valuation actuaries* is to determine the appropriate "just a little more" and validate the expected number of claims, which should be what was taken into account when setting the price of the insurance. Valuation actuaries also certify the reserves to government agencies.

Pricing Actuaries

Pricing actuaries are responsible for determining how much money a company is likely to make on a product. A product can be life insurance, which pays an agreed-upon sum to your beneficiary when you die, an annuity, which pays an agreed-upon sum every month as long as you live, or some form of health insurance, which covers the costs of medical care not paid for by a government plan, for example, dental and drug expenses. Pricing actuaries use the same assumptions as valuation actuaries when calculating the price of insurance to guarantee consistency and ensure that when valuation actuaries believe that they are adding a little extra to the reserves, they are really doing so. Pricing actuaries generally do not certify anything to anyone outside of the company.

Consulting Actuaries

Consulting actuaries spend a good deal of their time advising on defined benefit pension plans. These are trusts set up to fund tax-assisted retirement benefits at a rate spelled out in a legally certified document.

In the United States, senior consulting actuaries are usually members of the Conference of Consulting Actuaries (CCA). To become a *Member* of the CCA, candidates must have completed a minimum of 12 years of responsible actuarial work, defined as "work that requires knowledge and skill in solving actuarial problems." They must also be a Fellow or Associate of the Society of Actuaries or the Casualty Actuarial Society; or a Fellow of the Canadian Institute of Actuaries, the Faculty of Actuaries, or the Institute of Actuaries; or be enrolled with the Joint Board for the Enrollment of Actuaries (EA), thus having acquired the title of *Enrolled Actuary;* or be a *Member* of the American Academy of Actuaries, the *Asociacion Mexicana de Actuarios Consultores*, the *Asociacion Mexicana de Actuarios*, or the *Colegio Nacional de Actuarios*.

In the United States, for example, they must be Enrolled Actuaries to have signing authority. The Employee Retirement Income Security Act of 1974 specifies that they must therefore have participated in determining "that the methods and assumptions adopted in the procedures followed in actuarial services are appropriate in the light of all pertinent circumstances." The must also demonstrate *a thorough understanding of the principles and alternatives involved in such actuarial services.* Their actuarial experience must include involvement in "the valuation of the liabilities of pension plans, wherein the performance of such valuations requires the application of principles of life contingencies and compound interest in the determination, under one or more standard actuarial cost methods, of such of the following as may be appropriate in the particular case: normal cost, accrued liability, payment required to amortize a liability or other amount over a period of time, and actuarial gain or loss."

In the United Kingdom, Canada, and certain other countries, *Appointed Actuaries* play a role analogous to that of Enrolled Actuaries in the United States.

Pension Actuaries

Pension actuaries look at all members of a pension plan, their ages and salaries, and projects how much each would receive at retirement on average, given that some will terminate before retirement, some will get salary increases, and other such assumptions as to what might happen in the future. Then they look at the assets the pension plan has invested and determine, based on these two analyses, how much the plan's sponsor (usually an employer) needs to contribute to the plan each year. The pension actuary certifies that the contributions needed to fund the plan are adequate and qualify for a tax deduction for the sponsor.

Pension laws and pension regulations are country-specific. This is the one area in which the global mobility of actuaries is somewhat restricted. Special examinations must be passed in the country of employment to be a pension actuary. In the United States, pension actuaries must be Enrolled Actuaries to be eligible to perform government-related pension fund audits. Enrolled actuaries are also employed in the human resource departments of large companies.

Senior pension actuaries in the United States are usually also Fellows of the American Society of Pension Actuaries (ASPA), a designation that is awarded only after successful completion of a series of professional examinations. The basic examinations are those required to become an Enrolled Actuary, together with three additional ASPA examinations. A Fellow of the Society of Pension Actuaries must also be a Fellow or Associate of one of the following societies: the Society of Actuaries, the Casualty Actuarial Society, the Canadian Institute of Actuaries, the Faculty of Actuaries, and the Institute of Actuaries, or be a Member of the American Academy of Actuaries, the Asociacion Mexicana de Actuarios Consultores, the Asociacion Mexicana de Actuarios, or the Colegio Nacional de Actuarios.

Although you will see later in this chapter that the actuarial profession is globally mobile, pension actuaries in many countries must meet certain specific national certification standards.

Financial Actuaries

As the worlds of banking, insurance, and finance become more entwined, a new breed of actuary is emerging, known as a *financial actuary*. An advertisement for a senior financial actuary on the Internet describes one of the novel roles of actuaries in business. A company was looking for a senior financial actuary whose responsibilities included *developing, analyzing, and testing models of Internet credit card processing systems including product pricing, positioning, and consumer credit, in order to minimize risk and improve return on investment. You will communicate assumptions, results, and alternatives to staff and provide guidance in systems reengineering.* A suitable candidate was expected to have at least a Bachelor's degree in actuarial science, finance, mathematics, or a related field and be an Associate Actuary. In addition to appropriate experience, the candidate was expected to be an effective communicator, and creative thinking skills were essential. The company was looking for a self-starter with a strong statistical background and proven expertise in modeling techniques. Moreover, knowledge of the financial and management needs of an Internet real-time credit card processing company was expected.

What Does It Take to Become an Actuary?

Skills needed include mathematical ability, knowledge of and comfort with computers and computer modeling systems, and the ability to communicate complex topics in terms that customers can understand. Most actuarial positions require that you are at least an Associate of the Society of Actuaries, the Casualty Actuarial Society, the Canadian Institute of Actuaries, or have equivalent standing in an actuarial society of another country. If you are in a position that requires you to certify actuarial valuations and reports, you must usually be a Fellow of these societies.

Many actuaries in the United States are also members of the American Academy of Actuaries (*see* Appendix D), the *public policy, communications, and professionalism organization for all actuaries in the United States.* As Section 1 shows, actuaries in different countries belong to wide variety of national and international professional organizations that define and direct the future of the profession. At the international level, the International Association of Actuaries (*see* Appendix D) plays a central role in coordinating and advancing global actuarial interests.

1.2 Benefits and Rewards

In my many years as Director of an actuarial work/study program, I have interviewed hundreds of students who have chosen to be actuaries. They all have one thing in common—*they all love mathematics.* Here is what some of them, and some of their employers, have given as reasons for their career choice.

Q **Did you ever consider working in a non-actuarial field of applied mathematics (such as engineering) and if so, what tipped the scales in favor of an actuarial career?**

Twenty-five percent of all respondents to the survey said "No." There was no doubt in their minds that all they ever wanted to be was an actuary. The rest had considered other careers. Here is what some of them had to say.

Answer I am currently working in a non-actuarial field where strong mathematical and financial skills are highly valuable. Elements that persuaded me to leave the actuarial field were salary and opportunity at the top management level.

Answer Yes. Communications and media. But I found that an actuarial career provides a more secure job, a great work environment, a good reputation, excellent job opportunities, and diversification of tasks, especially at the entry level.

Answer I considered studying engineering. I decided to follow an actuarial career instead because I didn't like some subjects in engineering (chemistry) and because the business part of an actuary's job interested me.

Answer I considered studying engineering. But I like the fact that being an actuary means that you need to acquire knowledge not only in applied mathematics (the primary reason why we're all in this field), but also finance, economics, taxes, politics, and all those things make an actuarial career so interesting.

Answer I did consider many other fields, including engineering and medicine.

Answer I was thinking about studying mathematical economics. Learning more about the actuarial profession and how challenging it is made me change my mind, and I never regretted it.

Answer I initially was seriously considering going into pure and applied mathematics and even engineering, until I stumbled upon actuarial science. It was the combination of the high-level applied mathematics and business skills required in this field that finally tipped the scales in favor of an actuarial career. The fact that actuarial science led to a much more rounded career appealed to me immensely and really made all the difference.

Answer Not really—I've been gunning for this since Grade 10. The workload of an engineering student at university steered me away from that, and I didn't want to be a computer programmer for my entire life.

Answer Yes. Statistics. But I felt a training in actuarial mathematics was broader and that it would be easier to switch from actuarial mathematics to statistics than the other way around.

Answer Yes. I applied to engineering. I then chose to become an actuary because it is more of a big-picture profession than engineering.

To be an actuary you need to have a long-term vision. You need to understand trends in the economy and be able to predict where the economy will be moving in the future. The concepts and theories you learn in statistics train you to think critically, to analyze, and to recognize patterns and trends.

Engineering is a more technical field and is not as conceptual as actuarial mathematics and statistics. I'm a big-picture man, and I believe that in the actuarial profession you get to see a lot more of the picture sooner. I assume that this training can also be applied to other fields in the future. It is a way of thinking and goes beyond technical knowledge.

Answer I haven't so far, but I'd like to keep my options open. The biggest stumbling block would be to realize how much effort I've put into the SOA exams to become qualified as an actuary and then ask myself, "Do I really want to ditch everything I've done for my career, put more time into studying something else, and take a 30% drop in salary?"

Answer I thought of being a teacher, but decided I didn't have the patience for that and I was drawn to a rotational-program setting at an insurance company so that I could have the exam support and variety of rotations. I would consider being an adjunct college professor or teaching an exam review class.

Answer Yes. But I decided to go in actuarial science because it was something less well-known to me and I found that to be a real challenge.

Answer I did consider it, but the job market favored actuaries at the time.

1.3 A Typical Day

Let us take a look at a day in the life of an actuary. What are the typical tasks, and how does the day evolve? Obviously the answers depend on the nature of the company and the seniority of the actuary.

Here is what several actuaries and actuarial students had to say about this in the survey:

 Describe a typical day in the life of an actuary.

Answer Corporate stuff. Reserve valuations. Asset and liability management. Dynamic capital adequacy testing. Pricing.

Answer Reading, replying and sending e-mail, letters and phone-mail. Keeping in touch with the daily activities of my clients and current economic developments. Talking many times a day with the consultants I work with to keep track of the many projects going on and address issues if necessary. Producing reports of different kinds when a consultant has to meet with a client, depending on the client's needs and what the consultant wants to show them. Calculating performance figures from the different managers investing money for a client's fund, reviewing their historical performance and comparing it with a universe of funds and benchmarks. Following up on previous reports prepared for clients that need to be updated for the coming quarter. Verifying trust statements at the end of the month to make sure there are no discrepancies with the manager's data. Carrying out all kinds of calculations that are required by the consultants in their work with clients. Lots of teamwork.

Answer In the pension consulting industry, a typical day includes many phone calls with clients on subjects as varied as plan funding and investments, tax legislation, particular situation of given plan participants, union negotiations, benefit improvement, accounting treatment of pension plan, etc. Also, peer review of actuarial valuation results, planning and management of projects, business development, formal or informal training, internal or client meetings. It's rarely nine-to-five.

Answer A normal day in the life of an actuary at my level involves a lot of work with computers. Checking data, using programs to calculate liabilities for pension funds, personal calculations, all that can be done in a normal day. It is also not unusual to have training sessions on hot issues or new tools.

Answer I get to the office and check the e-mail and voice-mail messages. In the morning, I tend to work on projects until lunchtime and to contact my clients when problems arise. In the afternoon, I often have meetings with teams or clients, and I then keep on working on specific projects with different people.

Answer Consulting in group health insurance: technical work on actuarial valuation of post-retirement benefits. Core consulting: renewals, review of financial reports, benefits redesign, analysis of insurer's quotations on group insurance benefits. General advice to clients about current issues on group insurance benefits: phone calls, client meetings.

Answer For an actuarial intern, there is no such thing as a typical day. The tasks vary by intern and company but usually start with daily routine jobs such as updating data, checking the results of jobs run the previous day, and meeting with your supervisor. The remainder of the day is spent working on one or possibly several projects you've been assigned. Having junior status, an intern may work for more than one actuary and is often asked to run illustrations, compute premiums, search for data, make graphs, etc.

Answer There aren't too many typical days. Every day has some new wrinkle or challenge. Things that are done pretty much every day are working with spreadsheets to perform actuarial calculations, checking the reasonableness of the results of calculations (Are results reasonably consistent with your prior expectation of what the results should be?), communicating with both actuarial and non-actuarial co-workers in person, by phone, or by e-mail. And during exam season, studying for exams if you're still taking them.

Answer Here is an account of a typical day at the office. It's basically a 10-hour day:

8:00	Walk to the office.
8:30	Arrive at the office; read e-mail and news.
9:00	Finalize calculations for the report to client ABC; give directives to assistant.
10:30	Preparation for meeting with client A at 1 p.m.
12:00	Lunch with investment manager of firm.
13:00	Meeting with client A: presentation of the report submitted three days ago, discussions of the next steps and answer questions and recommendations.
14:30	Prepare memo to client A following meeting concerning issues raised.
15:00	Debriefing with manager for client.
15:15	Consult voice-mail and e-mail.
15:30	Peer review report for client B.
16:30	Help junior analyst with calculation program for client C.
17:00	Contact Trust D for trust statement figures as of mm.dd.yyyy.
17:05	Search for client E: Investment manager for an equity mandate.
17:55	Time entry for the day.
18:00	Go home (and study for actuarial exams!!!).

Answer Internship in a pension consulting firm: every day is different. Different projects and obstacles to overcome. Challenging. It's hard to adjust between school and work routines. When beginning an internship, I often find myself very restless because I am not used to sitting in one place for long. At school, I never sit in one place for more than an hour.

Answer I arrive at the office at 7:30 a.m. I am usually the first one there, and I enjoy the quiet time to go through my e-mail, do some deep thinking, and plan the day's work. I am in the corporate actuarial department. We set valuation policy for the company or, more accurately, develop our company's interpretation of the valuation standards set by regulators and the Canadian Institute of Actuaries. I am currently working on standards for applying the new *Consolidated Standards of Practice* to our valuation.

▶ *E-mail.* The first thing I do in the morning is to read my e-mail. I send an immediate response where I can, delete any notes where no further action is needed, store notes that form part of a discussion thread, and print anything that I need to spend more time on during the day.

▶ *Calendar.* Next, I check my calendar to see what meetings I have scheduled. Meetings can be a very significant portion of a working day, and if I have a memo or some other piece of work due that day, I need to do some short-term planning on how the work will get done on time. At this point I decide what I will actually do during the day. This will include meetings, project work, occasionally production work, and research.

Project work is a catchall phrase for deliverables that take longer than a day. This could include developing standards for valuation, implementing a new computer valuation system, collecting and coordinating data from different business units in support of a corporate decision. There always are one or two projects on the go that can absorb any available time in a working day not taken up by short-term requirements.

Production work is usually tied to a particular time of the month or year, and relates to reporting requirements of one kind or another. My production work is to examine and analyze the source of earnings reporting for the company. Research means reading some of the CIA or OSFI (Canadian Office of Superintendent of Financial Services) papers that have been prepared for our education. Most of this is directly relevant to my current job since my department interprets these papers for the company.

▶ *The Rest.* The rest of the day is spent doing the work I have planned. My door is open, and the plans I have laid out are easily derailed if something comes up with a higher priority, such as a question from upper management.

Answer In the case of a consultant: teamwork, meeting with client, calculations, revision of the calculations of others.

Answer A day in the life of an international benefits consulting actuary: consulting with clients of all sizes on a wide range of benefits-related issues including pension plan redesign, valuation, accounting, compensation and expatriate benefits coordination.

Answer This greatly depends on the level of responsibility held by the actuary, the size of the organization in which the actuary works, and the type of company: life versus P/C [Property and casualty], consultant versus insurer, and so on.

The answer also depends on the period in question. For example, year-end will keep corporate actuaries very busy, while no overtime may be required the rest of the year. In any event, a day in the life of this actuary (meaning me) goes something like this. Bear in mind the following background information: I currently work for a small P/C reinsurance company (five employees), with both actuarial and underwriting responsibilities.

▶ *Rating.* Most of the day I work with Microsoft Excel. My work involves rating (calculating reinsurance premiums), production of reports, or corporate functions such as calculating IBNR [incurred but not reported] loss reserves, doing DCAT [dynamic capital adequacy testing] work, and analyzing quarterly financial information. Knowledge of Microsoft Word and Microsoft Access is also required, since we often write memos and reports and all our data is stored in Access.

▶ *Lunch.* Lunch is usually spent at my desk, reading e-mail, newspapers or trade magazines, in order to stay abreast of current events in the world and in the insurance industry in general. From time to time, I may go out for a lunch meeting with a client or broker. Some travel is required from time to time.

▶ *The Day.* A typical day will see me coming in the office at 8:15 a.m. and leaving at 5:45 p.m.

Answer The day in the life of an actuary depends on a variety of circumstances: insurance versus consulting, life versus P/C, big company versus small, traditional role versus non-traditional role, and especially the line of business the actuary is involved with—and even that can vary from day-to-day!

Actuaries I have met have handled pricing, reporting, risk management, reinsurance, and corporate and industry issues. Some are in non-actuarial roles like underwriting and senior leadership positions. Some work on group benefits (long-term disability, short-term disability, life, accidental death and dismemberment), some work on annuity products (fixed and variable), some work on life products (term, variable universal life insurance, universal life insurance), some work in investments, etc.

I don't think that there is just one way to describe an actuary's day!

Answer In consulting: phone-mail, client calls with specific issues, tight deadlines, challenging work.

Answer In my experience, actuaries tend to while away their days solving problems. I believe a typical day for an actuary is made up of four basic functions.

► *Definition.* First, actuaries must carefully define the particular problem they are planning to tackle.

► *Research.* Actuaries must then research the problem. This research can range from using a library or the Internet to collect reference material to discussions with colleagues and coworkers.

► *Solution.* The third step involves the development, testing and documentation of a proposed solution.

► *Implementation.* In the final step, actuaries seek approval and implement their solutions. Some problems are frequent but simple. In that case, the actuaries can complete all of these steps for a number of problems in a single day. Often though, actuaries have more complex problems that must be prioritized and addressed in a disciplined fashion. Dealing with a mixture of short- and long-term problems can also be seen as just another daily problem that actuaries can expect to have to deal with.

Answer I am currently involved in client support for an actuarial software package. I am also involved in the training of the users of this system. It is used for pricing, valuation and other actuarial tasks. A typical day includes training of clients either in person, on the phone or through the Internet. Our clients are located mainly in Canada, the United States and Southeast Asia. I also answer e-mail from these clients regarding problems and questions they have with the system.

Answer I am an actuarial student working in the investment division of a company. I am the pricing actuary in my unit. I am responsible for pricing stable value products. I also work on product development. Once the market opens, I spend the first half of the day pricing cases. I am in close contact with the investment strategy group monitoring rate movements and analyzing certain risks. I am also in contact with sales, communicating rates throughout the day.

My days are not planned because most of the cases are sent overnight, so every morning I go to work prepared for another challenging day.

Answer As a consultant in the asset consulting services group, my day is best described as working on different client projects, meeting investment managers to learn more about their team and investment process, delegating and supervising junior staff, meeting with clients to present reports—and studying during the evening for completion of the SOA examinations.

1.4 Typical Responsibilities

How do beginning actuaries spend their time at work, and how do these activities change as an actuary's career advances?

Q What are some of the typical actuarial projects on which you have worked, and what specific knowledge and skills were required? Please give some illustrative examples.

Answer Union negotiations: they require strong analytical skills, a talent for multitasking, and the ability to work well under pressure.

Answer Typical projects that I have been involved with include the production of reports, writing, graphing, editing charts; project management (requires good planning); communication with consultants (requires knowledge of the clients I work with, knowledge of Word, PowerPoint, and Excel); returns calculations: requires knowledge of the database, and basic financial mathematics.

Answer Basic actuarial valuation: calculating the plan's liabilities from the data of the participants of the plan. Basic actuarial projects require rigor, methodology and planning. Preparation of accounting disclosure and calculation of pension expenses: knowledge of accounting rules and their application.

Answer I've worked on annual statements. A good knowledge of Microsoft Excel and pension plans was required. Being methodical and having good organizational and language skills are important. I've also worked on actuarial evaluations. The same skills and knowledge as for the statements were needed, plus a good knowledge of valuation software, as well as familiarity with the law and the valuation process (gain and loss, reconciliation, etc.).

Answer Typical projects that I have been involved with included:

▶ *Valuations.* Actuarial valuations: determination of the present value of annuity benefits taking into consideration demographic factors (mortality, termination, retirement, etc.).

▶ *Reports.* Financial reports: understanding how balance sheets work, statistical knowledge, analytical skills, credibility notions, software skills (Fortran, Microsoft Excel).

▶ *Computing.* Software skills are crucial in the actuarial field. A good grasp of Excel, AXIS, Microsoft Visual Basic for Applications, and even APL are a great advantage and are widely used in the field.

The main project I worked on consisted of reviewing and updating a computation made in the valuation system of an insurance company. My work was very specific and involved many calculations, running illustrations, and analyzing results.

▶ *Products.* I also needed to have a good knowledge of the various products sold and their specific details. For example, if my results seemed irregular, my first instinct was to look up the product I was examining for distinct features such as product design or recent repricing.

Answer Typical projects that I have been involved with included:

▶ *Reserves.* Calculating reserves: needed knowledge of actuarial mathematics (life contingencies, theory of interest) and general structure of reserves, as well as computer software. Knowledge of professional standards of practice is also needed.

▶ *Balances.* Calculating fund balances for retirement and investment products: actuarial knowledge of the theory of interest and computer software were essential. Knowledge of legislation regulating such products is also needed.

▶ *Design.* Design of insurance and investment products: Knowledge of the different mechanisms of insurance products, knowledge of different investment products, rules and regulations regarding those products, computer software, communications skills when working with others were essential.

Answer Typical projects that I have been involved with include actuarial valuations of pension plan liabilities; costing of plan benefit changes; pension expenses.

The skills required for these projects were basic technical skills: mathematical, actuarial and accounting rules, knowledge of internal valuation software, and knowledge of laws affecting pension plans.

I have also written reports to clients: letters, actuarial valuation reports, investment manager monitoring reports, etc.

The skills required for this type of work are the ability to translate complex issues into understandable words, writing skills, and communication skills.

Answer Reserve valuations, year-end and quarterly pricing, new products, modification of current products, DCAT [dynamic capital adequacy testing], business projections for the next five years, performed once a year, and MCCSR [minimum continuing capital and surplus requirements] calculations.

Answer Installation of new valuation systems, project management, ability to reconcile old and new valuation systems by results (actuarial, analytical abilities), ability to influence area over which you do not have direct control, sense for when an approximation is OK, pricing of retirement savings product, knowledge of corporate pricing targets and practices, software skills, ability to seek and accept input from producers, ability to reconcile conflicting priorities of sales, to deal with management and the corporate office, and the ability to build consensus.

Answer Most are in line with post-retirement benefit valuations. Specific knowledge required: applying discount and mortality data to benefits scheduled for a future date.

Answer Typical projects that I have been involved with included:

▶ *Valuations.* Pension plan valuations. They are needed to ensure that the retirement benefits promised to employees by their employers are available for their retirement life. A valuation calculates the value of those retirement benefit promises (pension liabilities) and compares them to the assets invested. A fully funded pension plan is a plan that currently has a level of assets sufficient to cover its pension liabilities.

Skills required: actuarial background to calculate the required values; programming skills to understand/program/run the system on which the liabilities are calculated; analytical skills to check, compare and compile results; up-to-date knowledge on current market and economic issues used to set and understand the assumptions used in the valuation.

▶ *Benefits.* Administering the benefits of expatriates working in various countries. Expatriates add another layer of complexity in benefits valuation since coordination is required between the host and home countries, as well as potential social security benefits earned in various countries.

Answer Typical projects that I have been involved with included:

▶ *DCAT.* A lot of work has been done recently on DCAT [dynamic capital adequacy testing]. Essentially, this is a financial model that projects the future financial condition of a company. The model can be deterministic or stochastic in nature. In my last three jobs, I have been involved to various degrees with this.

This type of project requires good understanding of accounting concepts (projection of balance sheet and investment income), investment concepts (calculation of market and book value of investments under various economic scenarios), financial concepts (calculation of corporate income tax), and statistical concepts (calculation of various probability scenarios). Developing appropriate business knowledge through finance, economics, investment and management courses can never be stressed enough.

▶ *Computing.* Other projects that I have been involved with usually only require a good understanding of actuarial concepts, acquired through coaching and through the examination process. Expertise with Excel is always a must. So are other computing skills: Microsoft Visual Basic and SAS being the most common ones.

Answer Typical projects that I have been involved with included:

- *Annuities.* I've worked on developing new annuity products and riders (i.e., product management: seeing an idea develop into a real product that is sold to contract holders). Within that process, I have worked with all business areas (compliance, legal, marketing, systems, etc.) to get an idea into a working product.
- *Ratemaking.* Other projects included setting the credited rates for our various fixed and variable annuity products.
- *Profitability.* I have worked with in-house actuarial software to examine profitability.
- *Verifications.* I have verified client illustrations to verify that what is being shown to a client for an annuity product's subaccount growth and death benefit calculations is accurate.
- *Reviews.* I have also done product reviews of our existing products to validate the pricing.
- *Economic Value.* I've worked on economic value—determining which areas of the company are contributing what value to our theoretical stock price.
- *Reinsurance.* I also worked in reinsurance where I dealt with reinsurance intermediaries and brokers to renew contracts. This also involved assessing the risk within our existing contracts.

Answer Renewal analysis (group insurance); financial statement analysis (group insurance); reserves analysis; post-retirement benefit valuation; report writing; various types of research; preparation of benefit statements; policies and booklets verification.

    Knowledge and skills: Computer knowledge (programming, Microsoft Word and Excel), communication skills (in French and English), writing skills, planning ability.

Answer Actuarial valuations (knowledge: methods for valuing liabilities); accounting procedures (knowledge: basic accounting); calculations (knowledge: laws and regulations, ability to draft reports, good reading comprehension); plan design (knowledge: industry trends).

Answer Typical projects that I have been involved with included:

- *Reserving.* Standard reserving projects. Involved applying various development techniques (mostly triangular methods) to estimate ultimate losses, determining liabilities on unearned premiums, discounting loss payments, and calculating provisions for adverse deviation.

Skills required: Analytical skills, technical knowledge of actuarial methods, common sense, familiarity with types of insurance, lines of business and coverages analyzed. An example might be the projection of asbestos and environmental liabilities arising from old, expired policies issued to commercial clients. Skills required: Strong analytical skills, problem solving, creativity, curve fitting, stochastic modeling, computer programming, knowledge of legal environment.

▶ *Pricing.* Determined indicated overall premium change, calculated required change in base rates and relativities for various rating variables.

Skills required: Analytical skills, technical knowledge and understanding of actuarial ratemaking techniques, curve fitting, good judgment.

▶ *Benefits.* Special Studies: impact of change in statutory benefits provided under accident benefits coverage for auto insurance.

Skills required: Strong analytical skills, problem solving, creativity, resourcefulness.

Answer An experience study is a typical but simple actuarial project. An actuary may be required to analyze an experience as often as each calendar month. To complete this type of project on such a frequent basis, the actuary generally keeps the process simple and may rely heavily on computer systems. This requires the actuary to know about probability and statistics as well as mortality table construction, finite mathematics, survival models and computers.

The calculation of an embedded value for a company or block of policies would be a complex problem and could take a dedicated team of actuaries a number of months. The actuary would have to know about the relevant methods for dealing with asset and liability data, selecting assumptions and reserving methods to apply to this data and implementing computer systems to translate all this into a simple range of values.

Answer Typical projects that I have been involved with included:

▶ *Illustration.* Programming of an illustration system. Skills required were programming, analysis, and client contact. Training of clients using the company's system. Skills required were knowledge of the system and training capabilities.

▶ *Reports.* Preparation of actuarial reports for court cases. Skills required were knowledge of laws and capabilities of writing the reports.

Answer Cash flow testing, economic value benchmarking, product development and pricing. Helpful courses: Life contingencies, theory of interest, knowledge of fixed income securities.

Answer An actuarial background is not a prerequisite to work in the asset consulting services group. Some of my colleagues have a finance background. In fact, the projects I work on are not purely actuarial projects.

Answer Typical projects that I have been involved with included the following:

- ▶ *Assets.* How should the assets of a pension plan be invested? These projects are mostly worked on by actuarial people. They require a knowledge of both the liability and assets sides of a pension plan: demographics, financial results, investment markets, etc.
- ▶ *Investment.* Review the pension plan (statement of investment policy and procedures).
- ▶ *Management.* How to implement an investment policy, how many investment managers to assign to each asset class, what kind of investment managers to select (large/midsize/small capitalization, value/growth/core investment style).
- ▶ *Personnel.* Selection of investment managers for each asset class (Canadian equities, US equities, international or foreign equities, fixed income, etc.). A management structure and the manager selection require a good knowledge of the institutional investment market; you need to know the players, their investment process and style, and so on.
- ▶ *Monitoring.* Monitor the investment performance of each manager. It requires a good knowledge of their style as well as how the markets are performing in order to really understand their numbers and be able to explain their performance to clients. You need to know the team players in order to monitor any changes and turnover of people.
- ▶ *Mandates.* Put in place the appropriate paper documents between the pension plan and the investment managers.
- ▶ *Records.* Defined contribution record keeper selection.
- ▶ *Options.* Defined contribution investment options selection.

Entry-level Jobs

Q What are the responsibilities of new employees in actuarial entry positions in your company, and what are their typical tasks and salary ranges?

Answer Preparation of reports, letters, documentations, filing for clients. Technical knowledge to accomplish the work. Communicating effectively with consultants by phone and mail. Quality of the work done (as accurate as possible). Passing exams. Salaries range from $35,000 to $50,000 CAD.

Answer Entry level employees will normally work on plan participant data that will be used for actuarial valuation purposes. Preparation of annual statements. Preparation of various worksheets, projection of calculations, etc.

Answer With three actuarial exams, the starting salaries are about $45,000 CAD. At the starting level, actuaries are more technical experts. They are in charge of the computer work and getting to know their clients' plans, laws, etc.

Answer Technical analysis, renewal, financial reports, database statistics on current topics. Salaries are between $30,000 and $35,000 USD.

Answer New employees in a company are usually actuarial assistants or analysts, and must focus on learning all they can about the operations of the company. Knowing and understanding how and what the company does are crucial. Responsibilities consist of testing products, running various scenarios, researching and observing trends in the market and how the company fares, gathering data, computing premiums, etc. As experience is gained, more responsibilities are given. Salaries will vary according to the number of professional exams, but should rank between $30,000 and $35,000 USD.

Answer New employees tend to work on specific projects or may be assigned tasks that are periodic in nature. Difficult to describe or list specific tasks since they are usually company-, department-, or manager-specific. New students are expected to be able to develop their skill at judging reasonableness of results, using the experience they gained by working with their managers. Students may be asked to summarize results from reserve calculations to see if they were done correctly. They may also be asked to participate in research studies in which they may perform many data-massaging exercises. Salaries for a beginning student in the United States seem to be between $45,000 and $50,000 USD.

Answer Responsibilities: Limited responsibilities. New employees are expected to:

▶ Understand what is asked of them (by asking questions, taking notes, etc.) and return what is expected (quality job, list of questions that arose while doing a job, etc.).

▶ Acquire basic technical skills, adapt to and learn internally used software, procedures, etc.

- ▶ Participate in various portions of projects, supported by more senior employees.
- ▶ Typical tasks: Compile, clean, and analyze data. Programming required for actuarial valuations. Help in preparing reports (stats, other calculations, etc.).
- ▶ Salaries: Vary by province, city and country. In Montreal, salaries would be between $30,000 and $35,000 CAD, depending on the number of actuarial exams, company size, etc.

Answer Responsibilities involve number-crunching. By that I mean doing all basic calculations involved in a project. From calculating a projected cost to a simple projection of a cost. In my case, with a bit more than a year of experience, I often need to value future benefits. This involves calculating the present value of future benefits for all employees of a given client.

Answer Entry-level positions are usually actuarial analyst positions. Candidates are expected to be actively completing SOA exams and usually have passed the first few courses when hired. Responsibilities: working closely with senior analysts and junior consultants on a wide range of client projects. Project responsibilities usually include analyzing data and programming, calculating pension plan liabilities, compiling and analyzing actuarial valuation results, answering various day-to-day client questions relating to pension plans, special projects. I'm not sure what starting salaries would be. In Toronto, probably somewhere around $55,000 CAD, depending on the number of exams passed.

Answer My company does not have entry-level positions. However, my previous company (a P/C insurance company) does have such positions. Typically, a starting salary will depend on the number of exams passed. University graduates with two exams could be making about $45,000 CAD in Toronto.

Salary ranges vary considerably depending on which city you live in. Responsibilities would include the production of routine reports, the preparation of ratemaking or reserving, Microsoft Excel spreadsheets for analysis by more senior actuaries (and eventually some analysis with the help of the senior actuary), programming, and data entry. The quality of individuals will most often dictate how quickly their salaries rise, and how quickly more responsibilities are assigned to them.

Answer Entry-level students can play a role throughout the company in a variety of positions. Typical starting salaries would probably be about $50,000 CAD, assuming the individual had passed one exam. Students can work in financial reporting roles doing the monthly, quarterly, and annual output, and in product development roles assisting more experienced actuaries. In our company, an entry-level student might be placed wherever assistance is needed. Entry-level students are expected to build upon their technical and communication skills as well as pass exams.

Answer Responsibilities and tasks: Gradually communicate with clients, write reports and letters, carry out calculations and do research, prepare internal presentations, and write articles on "hot" subjects.

Answer Basic valuation work: individual calculations.

Answer Responsibilities: analyze and price or reserve casualty, life, or health insurance products. Salaries range from $42,000 (base salary for candidate with no prior internship, no actuarial exam, weak grade point average) to $54,000 CAD (candidate with advanced degree, prior internship, strong GPA, and three actuarial exams: $3,000 CAD per exam).

Answer Programming in Microsoft Visual Basic. Training of clients.

Intermediate-level Jobs

Q What are the responsibilities of employees in intermediate actuarial positions in your company, and what are their typical tasks and salary ranges?

Answer At this level, an actuarial employee will be responsible for interfacing with clients on a daily basis, as well as peer review, the management of projects, and the supervision of junior staff.

Answer More consulting: meetings with clients, providing advice, preparing of documents for presentation to clients, reviewing of more junior technical work. Salaries range between $45,000 and $50,000 USD.

Answer Intermediate actuarial employees are usually assigned specific projects requiring good problem-solving skills. For example, they could be asked to come up with a new way of computing certain data that are presently time-and-cost-consuming, or may be asked to design a new approach for calculating reserves for a new product with non-traditional features. Salaries may range from $35,000 to $40,000 USD.

Answer Intermediate actuarial employees should not only be able to handle routine tasks and mathematical model building that an entry-level employee would need to do, but should also be able to significantly modify or redesign models. They are expected to have a more complete understanding of the industry practices and regulations, and be able to use their judgment in applying these standards to working situations. Decision-making ability should be more developed. Salaries may range from $60,000 to $75,000 CAD. Similar ranges in USD apply to students in the United States. New FSAs usually start at about $75,000 to $80,000 CAD.

Answer Intermediate actuarial employees have broader responsibilities. They are expected to be able to lead small to medium-sized projects through all the steps and train junior employees. Typical tasks included the preparation of reports, peer review of calculations and programs. Salary ranges vary by province, city, and country. In Montreal, salaries range from $40,000 to $50,000 CAD, depending on the number of actuarial exams, company size, etc.

Answer Assistant actuaries (first level after becoming an FSA) should be able to write technical memos presenting assumptions, data, discussions and conclusions for an audience of actuaries. For example, they should be able to write a note on how the investment assumption for a business unit was developed, describing the asset classes, the starting yield curve, the development of PfADs [provisions for adverse deviations], the reinvestment assumptions, and any planned changes to the investment policy. They could also be valuation managers (persons who actually run the valuation programs and determine the reserves, under the management of a valuation actuary).

Answer Intermediate actuarial positions in my view would be for analysts with three to five years of experience, near qualification as FSAs, and are usually ASAs.

Salaries in Toronto probably are between $65,000 and $80,000 CAD, depending on the number of exams and performance.

Responsibilities include working closely with junior analysts and consultants on a wide range of client projects: coordinating and managing projects, delegating project tasks to junior analysts and preparing final projects for presentation to clients. They are involved in day-to-day client queries and projects, and participate in relevant client meetings and discussions.

Answer My current company does not employ intermediate actuaries. In my previous company, this type of position was held by people with five to seven P/C exams (out of nine), and/or four to five years of experience. Salaries depend on years of experience and successful exams, and range from $40,000 to $50,000 USD, or more. Intermediate actuaries tend to be responsible for specific actuarial projects (rate review for a given product and province, review of the IBNR [incurred but not reported] loss reserves, planning premiums and loss ratios for the budget, monthly review of results), subject to the supervision of a manager (typically, a Director or Vice-President). They may or may not have the help of a junior actuary. They also tend to be the experts assigned to company-wide projects, whether the project is an IT [information technology] (new rating engine) or business project (review of claims reserving practice by claims department).

Answer Most intermediate students (having passed three to six exams) probably make around $60,000 to $80,000 CAD. They work in any of the various areas of the company and are expected to be further honing their technical and, especially, communication skills. Typical tasks include a more advanced role in various areas of the company since they are expected to understand the corporate structure and have at least one to two years of experience behind them. Often they are encouraged to be managers of summer interns to get some management experience.

Answer Typical activities of intermediate actuaries involve communication with clients, preparation of reports, verification of calculations, sales, presentation to clients, training of other employees, and billing.

Answer Their activities include the delegating of work to junior staff, checking their work, and dealing with clients.

Answer Same as for entry-level positions. Salaries range from $50,000 to $125,000 USD.

Progression of Responsibilities

Q What are typical SOA and CAS career paths and where should successful actuaries or actuarial students be at age 20, 25, 30, 35, 40, 45 in (a) SOA, (b) CAS?

Answer At 20: at school. At 25: junior consultant. At 30: pre-senior consultant. At 35: senior consultant. At 40: advanced senior consultant. At 45: responsible for major clients, line of business, direction, etc.

Answer I don't think there is such a thing as a typical career path. For SOA, actuaries should have a firm client relationship with clients by the time they're 30. At 45, they should be established as client managers, responsible for high-level work and the relationship with clients.

Answer At 25: SOA graduate with the first three or four exams. Then continue to write exams while working and finish before 30. At that age, actuaries should be familiar with the technical concepts and begin to be relatively autonomous in establishing what needs to be done on different projects. At 30, they should be able to review the work of junior students and have their own clients. At 35, they should be senior consultants.

Answer At 20: finishing an undergraduate degree in statistics or actuarial sciences and have at least written the exam. At 25: be at about Courses 4 or 5, and have spent one or two years as an actuarial assistant. At 30: have completed all courses and have gathered five to eight years of experience in one or more companies and hold an Assistant Manager's position. At 35: manager or director. At 40: permanent senior position, secure and confident in the position they are holding.

Answer In my opinion, this should be stated in terms of duration from when the first exam is attempted, rather than by age. People get into the field at different ages and different places; they have different average ages upon graduation from college. Thus, it is not uncommon for someone to get their FSA prior to age 25 in the United States, whereas it is less common in Ontario because Ontario students graduate from university when they are between 23 and 24, instead of 21 or 22. I have met people who didn't start taking exams until their 30s because they switched careers. Most students should get their Fellowship about 8 to 10 years after they started taking exams. The average age of new FSAs is usually in the mid-30s, although the SOA wants to reduce the exam travel-time and, indirectly, the average age of new FSAs. They want to do this, but I doubt it will happen.

Answer I will cover only the SOA exams. At 20: start the exams if you want to be an actuary. At 25, you should have completed Courses 1 through 4. During the first few years of your actuarial career, you will be a junior actuarial analyst. At 30, you should at least be an ASA. You will be a senior actuarial analyst or junior consultant. At 35, you should be an FSA or have decided whether you want to continue writing exams. Life consists of more than SOA exams!!! You should be an intermediate consultant. At 40, you should be a senior consultant.

Answer I would like to have finished all of my SOA exams by the age of 24. After a three-year university program, a solid goal is to have passed four exams.

Answer At 20, you should be in university and have started the first two or three exams. At 25, you should ideally have finished your exams and should be waiting for the completion of your PD [professional development] requirement credits. At 30, you should have two or three persons to whom you delegate work and start helping them build their knowledge. At 35, you should focus on networking and meeting people, start bringing clients to your consulting firm and maintain relationships with existing clients. At 40, you should probably be at the peak of your responsibilities.

Answer At 20: actuarial student. At 25: senior actuarial student. At 30: FSA. At 35: Associate Actuary. At 40: Assistant Vice-President. At 45: Assistant Vice-President or Vice-President.

Answer From the SOA point of view: At 20: in university. At 25: starting out, passing exams, gaining experience at an actuarial firm, deciding on insurance versus consulting, SOA versus CAS, is or is close to being an ASA. At 30: is or is close to being an FSA, settling into the actuarial field with preference for insurance or consulting, SOA or CAS. At 35: twelve or more years of experience. Consultant level with expertise in a preferred field. Providing valuable advice to clients on a wide range of client issues, and a good source of intellectual capital for peers.

Answer I will answer this question from the CAS perspective.

▶ *Student.* Typically, at 20, you will still be in university. You will hopefully get some summer work experience, if not working for an insurance company, at least getting some exposure to the office world. You should be planning to write a few actuarial exams while in university to show prospective employers your willingness to write exams, and your capacity for writing them successfully.

▶ *Intermediate.* At 25, you should be making the transition from entry-level to intermediate actuary. You should have written several exams by now, including basic ratemaking and reserving (although not necessarily passing them), which will prove invaluable in the new responsibilities being handed to you.

▶ *Associate.* At 30, you should be an Associate, even a Fellow if you are one of the more gifted. This is the point in your actuarial career where you are handed management responsibilities. Although everyone wants to be a manager, very few understand what is involved. If working for a good company, the actuary will have been sent to some form of management and other business-training seminar. But the very motivated individuals will not rely on the company, and will read up on these subjects at home.

▶ *Vice-President.* At 35, most CAS Fellows are Vice-President or its equivalent such as partner in a consulting firm (at least, in Canada). Responsibilities start shifting from the pure actuarial areas to the areas of company management and client management.

▶ *Career Peak.* At 40 and 45, your level of responsibilities will slowly increase, but essentially, things will remain the same until retirement.

Answer I cannot respond with respect to the CAS, so the answers below are with respect to the SOA.

- ▶ *College.* At 20: taking college courses towards a mathematics degree or actuarial degree. Investigating internship opportunities. Planning to take one or two exams before graduation.
- ▶ *Work.* At 25: working at a company with one to three years of experience. Have passed two or more exams.
- ▶ *Almost FSA.* At 30: working for a company with 5 to 8 years experience. Be close to attaining FSA if not already an FSA.
- ▶ *FSA.* At 35: have the FSA designation, have 10 to 13 years of experience, have a staff working for you, have a more prominent position and be out of the "rotational" student program. Know your specific area of interest or track the one you want to pursue in depth.
- ▶ *Leadership.* At 40 to 45: have a significant leadership role within the company and a staff working with you. Be accessible to newer students in the program who want advice.

Answer In my answer, I will focus on a CAS career. At 20: nice to have passed at least one exam. At 25: two years of experience and at least four exams. At 27: five exams. Very good candidates will have access to a managerial position. At 30: will probably be a Fellow by this age—if not, no problem, but focus on finishing the exams. At 30 to 35: outstanding candidates will have access to senior management positions. Over the first ten to fifteen years of an actuary's career, it is not uncommon for a person to have worked for several employers.

Answer I will describe a typical SOA path. At 20: junior staff in consulting firms or insurance companies. At 25: almost a consultant. At 30 and above: senior consultant for clients and relationship manager.

Answer Here is a typical CAS career path:

- ▶ *School.* At 20, students are still in school, completing their Bachelor's degree (or Master's degree, even though it is not required in the actuarial field). While in school, students generally start taking actuarial exams. A successful student should have passed the first two exams before graduation and should have had at least two internships related to the actuarial profession or insurance industry.
- ▶ *Analyst.* At 25, students should have a year or two of experience and be well established as an actuarial analyst. At that point, a successful student would have passed 5 or 6 actuarial exams.

▶ *Almost Fellow.* At 30, actuaries have typically been exposed to various aspects of the actuarial profession and have expanded their experience to pricing and reserving different lines of insurance. They also have analysts reporting to them and should be close to obtaining their Fellowship (if not done already).

▶ *Vice-President.* By 35, actuaries should definitely have their Fellowship and be in a management position (either as a senior consultant in a consulting firm or a Vice-President, or Assistant Vice-President, in an insurance company).

▶ *Partner.* At 40, a successful actuary would be a partner in a consulting firm or an officer in an insurance company.

▶ *Retired.* At 45, a *very* successful actuary would retire.

Answer At 20: in university. At 25: actuarial student, 35 hours per week on the job and 40 hours per week studying. At 30: new Fellow, supervisor or manager of a few actuarial students and clerks, or a highly technical position without direct reports. At 35 and above: continually increasing responsibility, demonstrated by increased staff and budget or required technical knowledge.

Answer I am not aware of requirements for CAS. At 20: in university writing exams. At 25: out of university in a junior level position. At 30: almost done with the exams and with some supervisory responsibilities, changing departments on a biannual basis. At 35: done with the exams and with more supervisory responsibilities. At 40 and 45: same as 30 and 35, but settled into a department.

1.5 Mathematical Skills

Here is what the respondents to the survey had to say about the basic mathematical knowledge they require in their daily work. They also commented on the connection between theory and practice. What links are there between the actuarial examinations and their required working knowledge of mathematics, finance, economics, and other special subjects such as risk theory, loss modeling, and stochastic methods?

Q **What general mathematical competencies are required by an actuary? Give some examples and relate them to the SOA or CAS examinations.**

Answer Return on asset calculation (Course 2), retirement plan methodology and characteristics (Course 5), statistics related to risk (Courses 1 and 3), pretty much all of Course 6 for me (as a junior) in asset management, basic financial mathematics (Course 2).

Answer Problem-solving, but this has nothing to do with any university course or actuarial exam.

Answer Calculus is needed for the first actuarial exam. Financial mathematics is very useful in the day-to-day work as well as for the exams (tested on more than one exam). The whole of actuarial theory is based on statistics, so it is, of course, a required competency.

Answer For the first exam, you need a lot of basic probability and calculus competency. The second exam is more about financial mathematics, macro- and microeconomics, and finance. The general mathematical competencies required for this exam are mostly integrals and derivatives. After that, you will always be using a variety of mathematical competencies (again basic probability, integrals and derivatives), but they will become more specific.

Answer Course 1 deals with basic probability and calculus competencies. An ability to deal with them and apply them to actuarial problems is crucial. In general, a deep understanding and competency in probability and statistics is essential to passing SOA and CAS examinations since they are the foundation of actuarial mathematics. A strong background in statistics is necessary.

Answer Knowledge of probabilities and statistical distributions, life contingencies, theory of interest, calculus, geometric series. The calculus is often tested via continuous distribution functions where integration of a function is required (Course 1). Probabilities of people living and dying are combined with geometric series to create the mathematics of insurance and annuities (Course 3). The theory of interest is used for the principles of interest discounting and accumulation (Course 2).

Answer Theory of interest, life contingencies.

Answer Theory of interest is a must (time value of money). Probabilities are also very important.

Answer Well, everything that's mathematical in the exam syllabus. Plain and simple!

Answer Competency in calculus, statistics, algebra, probability is essential, especially for the early exams.

Answer Actuarial mathematics such as life expectancies, survival models and projections, annuity factors—regression analysis—e.g., calculating trends, building models, etc.
    Calculus: background used in most programs and models.
    Statistics: always needed to calculate averages, medians, quartiles, etc.

Answer This is a difficult question. The answer also depends on the level of sophistication reached in the various companies. P/C companies in Canada are small and not a lot of complex mathematical models are built. I know of one or two companies working on that front, and they have hired a person with a Master's degree in statistics to do the work. However, these people are supervised by actuaries. Advance knowledge of calculus, statistics, theory of interest, life contingencies, and loss distributions are generally required to pass the first four exams. Past that point, at least on the P/C side, mathematical competency almost boils down to being able to add and multiply. Basic knowledge of the above is all that is required. And as I said, I find the same is true for our day-to-day life at work.

Answer CAS: only basic mathematical competencies are required. Regression and modeling may be beneficial, but are not a must. It is a common mistake to believe that extensive knowledge of mathematics is required to be an actuary. However, one must like to work with numbers to enjoy being an actuary.

Answer Basic mathematical skills needed. The examinations helpful for a career are the ones that discuss the different methods for valuing liabilities, accounting, and finance.

Answer Calculus, probability and statistics.

Q Why do actuaries need calculus? Please give examples and relate them to the SOA and CAS examinations.

Answer Rarely used so far in my career, and if I happen to need a concept from calculus, I can easily find someone in the office who will be sharper than me on that subject. The more I advance, the less I see a calculus background as being useful at work. But I can understand that it is a great mathematics background to have as an actuary.

Answer They don't need it for most of their day-to-day work.

Answer Actuaries study things that change as part of their daily work. Calculus is the mathematical construct that is used to quantify, measure and discuss how things change. I don't think anybody who truly understands change should have problems with calculus. People who have difficulty with calculus will probably lack the problem solving skills that are required of an actuary.

Answer Mostly for the exams. So far, I've never used it in my job.

Answer I don't believe calculus is actually used directly in the everyday life of an actuary, but it is a mathematical concept that needs to be understood by anyone who is said to be an expert in mathematics.

Answer They don't.

Answer From my point of view, calculus is only helpful for the first actuarial examination. After that, you will only use simple applications.

Answer Calculus is a basic tool used in probability that must be mastered. Questions arising in Course 1 for example, will deal with those competencies. Also, Course 1 will specifically ask calculus questions. Calculus is also a basic tool used in actuarial mathematics. In Course 3, for example, it is crucial to have a good grasp of calculus to successfully pass this course.

Answer Probability of paying a death benefit on any day required integration over a continuous distribution function—which is calculus. Also, trend analysis uses predicted rates of change, which is calculus. This mostly crops up on examinations in Courses 1 through 4.

Answer Knowing how to integrate or estimation using sums is the basis of most actuarial valuation formulas. Integrating is also the basis of modeling (e.g. using the normal or lognormal distributions).

Answer I do not find any direct application, although it could have helped in providing me tools for analysis and workout for my brain.

Answer I love calculus; I always have. I think that knowledge of the properties of basic functions, continuity and multidimensional spaces is essential food for thought. I think that the practical applications are limited. Basic calculus is tested in the Course 1 exam. Continuous life insurance premiums, coverage and annuities are dealt with in life contingencies, Course 3, but I doubt that any of this is used in practice. It is still good conceptual training.

Answer Concepts like "rate of change when delta *t* is minimum." Things such as force of mortality. You need a calculus background to grasp exactly what it means. Personally, I think probability and statistics is much more important.

Answer To evaluate continuous probability density functions, to evaluate continuous mortality functions.

Answer Needed in order to understand the underlying models and processes. Although most actuaries don't sit around to derive and integrate all day, calculus is required to understand the underlying actuarial formulas, calculations, processes, etc. Computers do most of the work, but calculus is a basic building block. The curriculum on the mathematical SOA exams is always more technical than the skills you'll ever need in real life.

Answer I'm not sure there is a great need for calculus in our day-to-day job. However, calculus will help form a problem solving mind set, I find. I have used calculus, personally, to calculate, for example, the average earning period for our unearned premium (simple matter of integral). This is a really basic Course 1 question.

Answer They do not in their day-to-day work. However, taking calculus is part of having a general knowledge about mathematics. I would not discontinue calculus courses—or any other mathematics subject for that matter—because they are not used in our actuarial day-to-day work. As far as the exams are concerned, knowledge about calculus is needed to be able to answer the questions. That's it.

Answer Personally, no.

Answer Understanding the general formulas.

In addition to calculus, Course 1 covers basic probability and statistics. Here is what working actuaries and actuarial students had to say about their view of the importance these topics:

Q Why do actuaries need probability and statistics? Please give examples and relate them to the SOA or CAS examinations.

Answer Laws of probability and statistics are useful in my work, but only basic concepts are needed on a daily basis. These concepts must be very well understood. For the rest, I consult books when necessary.

Answer To understand the concepts of risks, management of outcomes and impact on plan liabilities.

Answer You can't be an actuary if you don't understand statistics. I use it every day at work. Not necessarily the way I learned it in school, but at least the basic principles. It is used in actuarial valuation (with decrement tables, annuities, etc.) and in many other day-to-day actuarial tasks. It is also required for the exams (directly in the first exam and as part of actuarial theory in the others).

Answer The main purpose of actuarial mathematics is to calculate risk, and the only way to do this is through probability and statistics. For an insurer, the only way to figure out how much to charge his customers is by calculating how much they are more likely to claim.

Answer Virtually every business problem the life actuary deals with involves the assessment of risk, i.e., the value of a future event contingent on assumed probabilities. Being comfortable with this concept is essential for the daily work; actually applying advanced statistical concepts is much less of a requirement, although the opportunities to do so are increasing.

Answer Probabilities are a big part of the first exam and statistics, big part of the third exam. If you chose to work in the CAS field, statistics will certainly be a bigger part of your work and study than in the SOA field.

Answer Probabilities and statistics are the root of actuarial science. They are essential to the actuary and must be mastered. To understand and grasp actuarial mathematics subjects such as life contingencies, it is necessary to grasp the basics of probabilities. In Courses 1 and 3, those skills will be tested.

Answer Probabilities of events occurring or not occurring are the backbone of actuarial science. Probability of death, of a car wreck, probability of continued survival. Use of probabilities and statistical distributions can appear on any exam, although mostly in Courses 1 through 4 and Course 7.

Answer Most events and risks evaluated by an actuary are contingencies that can generally be assumed to follow a probability distribution. Actuaries also calculates probabilities (event to occur, having a negative return), expected values (rate of return, age at death, etc.) and volatilities (rate of return, sensitivity of the liabilities).

Answer Just to understand the basics of an actuarial valuation, you need a strong knowledge of probabilities. Once you get that, you have the power to modify the contents of a valuation, or to solve a totally new problem.

Answer Ah, it relates to the calculus question. Everything that contains the word "expected" relates to probability and this is the core of actuarial science. "What is my expected loss or expected profits on this block of business?"

Answer Mortality tables are themselves probability distributions, aren't they? Statistics helps us assess the mathematical validity of the tables by means of confidence intervals, and guides us in determining how much data to collect.

Answer Probability and statistics are again very basic building blocks needed to analyze data, build models, etc., in the projects assigned at work. Again, the curriculum on the mathematical SOA exams is always more technical than the skills you'll ever need in real-life.

Answer Some form of probability or statistics is used on a monthly, if not weekly, basis. Examples would include fitting a curve and testing its fit for calculating trends; calculating the probability of an event; calculating the standard deviation of a series of observations; performing a Monte Carlo simulation; etc. Most of this is covered in the first four exams of the CAS.

Answer I work mostly in pension. In that field of practice, probability and statistics are very important since most of the calculations are based on probabilities. Examples: the probability of someone surviving to retirement, the probability of dying a few years after retirement, the probability of someone leaving the workforce before retirement.

Answer Courses 1 and 4 have the majority of the statistics problems. It is important to have this background—probably more so in P/C. Credibility of past experience often plays a role, as does frequency and severity. Also, within risk management, stochastic scenarios and the distribution shape are important to consider.

Answer Probability and statistics may be used from time to time on the CAS side to estimate the price of new products (we have no data for those). For instance, in estimating how much a credit card company should charge to provide "delayed baggage" insurance, an actuary could answer the following questions (and then estimate the cost of providing this "coverage"): What is the probability that the baggage be delayed? What is the probability that cardholders with this "coverage" will be aware that they have it, and will then use it? What is the expected value of the loss, and what is the impact on the cost of providing the service of various "limits of coverage?" I find it very hard to relate this to examinations (I wrote them a while ago).

Answer In retirement consultation, very useful for valuating a pension plan.

Answer Probability of decrements.

Q Why do actuaries need the theory of interest? Please give examples and relate them to the SOA or CAS examinations.

Answer Very important for what I am doing, time value of money is a key concept, actuarial present values, rate of return formulas, amortization tables, etc., are all concepts that I have to play with very often in my work even if the way I work with them is different from an examination in Course 2, for example. Excel is used a lot in playing with these concepts.

Answer This is the basic element of the calculation of today's value of any future payment of one dollar. It is the cornerstone of our field.

Answer It is essential for the calculation of annuities and the understanding of the time value of money. For example, we use it when we calculate things payable at retirement with money accumulated today, or when we want to know what is the value of a pension fund today considering what the membership of the fund might be in the future. Again, it is tested directly in one of the first exams and comes back indirectly in the others.

Answer When dealing with a client, we are looking at the overall result of the company, and this includes investment income, future claims, future revenues, etc. The theory of interest is crucial when it comes the time to take those amounts into consideration. It would not be right to use an amount that will be obtained in 10 years, and this is where discounting comes in. The whole point of theory of interest is to calculate the company's financial situation at a certain point in time.

Answer Similar to the previous question, virtually every business problem the life actuary deals with involves the assessment of risk, i.e., the value of a future event contingent on assumed probabilities. The present value of a future event requires the application of the theory of interest.

Answer Theory of interest is the basic of many actuarial mathematics and finance concepts. This material teaches you the value of money in the time. This has many applications in the every day and in the work life. It is also a big part of the second exam.

Answer Present value of annuities.

Answer The theory of interest is needed to understand the basics of actuarial mathematics. The simple concepts of Present Value and Annuities are present, introduced and explained in details in theory of interest, and are everywhere in actuarial science. In Course 2, these skills are tested.

Answer Time value of money (i.e., accumulation and discounting) and understanding the basic structure of a bond are very important for calculating reserves, premiums. They are crucial for asset-liability management. Courses 2 through 8 use these concepts.

Answer Basis for discounting future value of loss, benefits, etc. Also used in projecting figures in the future.

Answer That is *the* required course. If you don't understand this one, you may as well forget an actuarial career.

Answer The theory of interest is crucial. The time value of money is one of the underlying principles of the insurance industry, only insurance takes it one step further by applying statistics.

Answer Probably not all that necessary now that most work is done on computers using interest vectors.

Answer The theory of interest is one of the essential building blocks of actuarial mathematics. It is needed to define present and future values, for example. Concepts such as calculating present values of bonds and calculating loan payments and outstanding mortgage values involve the theory of interest.

Answer Actuaries in property and casualty insurance are constantly discounting future streams of payments to calculate present values. They also need to understand annuities since they are sometimes used in the claims settlement process. Beyond this, it is not being used too much.

Answer In pension, the theory of interest is an important subject. The payment stream after retirement is based on mortality and interest. It is also needed to project ahead or discount employee contributions.

Answer Interest theory and time value of money are extremely important in any investment-product context (Course 2 of the SOA exams). Reviewing cash flows, profitability, and understanding gain/loss scenarios all hinge on the theory of interest. It is particularly important for actuaries in the investment field.

Answer CAS only: present and future value calculations (investment of insurance funds and discounting of loss reserves). Some annuity calculations.

Answer For valuating pension plans we need the concept of present value.

Answer Calculation of present value of future stream of payments.

Q Why do actuaries need the mathematics of finance? Please give examples and relate them to the SOA or CAS examinations.

Answer Finance is a big part of second exam.

Answer Mathematics of finance is also needed to understand the basics of actuarial mathematics. In Course 2, an extensive and deep understanding of finance is needed. This knowledge and skill will also be used in the workplace. Often, an actuary will be asked to do some financial analysis. A good basis in mathematics of finance is necessary for a good actuary.

Answer Actuaries need to understand assets as well as liabilities in order to properly set reserves and premium and dividend rates. Actuaries now need to understand both sides of the balance sheet to do their job correctly. Courses 5 through 8 (SOA) really hit on this.

Answer Needed when working on the asset side of a pension plan.

Answer The Course 2 exam. Also, insurance products relate very closely to the time value of money and finance.

Answer Financial mathematics is used when valuing pension assets. Also, a basic knowledge of financial markets is always useful when dealing with clients and in devising models. Investors and their advisors are becoming more and more informed, leading to more sophisticated market developments, products and services. As an actuary, and in most cases, at least indirectly affected by financial markets, a basic knowledge of financial mathematics is highly recommended. Course 6 of the SOA examinations is almost entirely based on financial mathematics. Although probably more technical than most actuaries will ever need, it provides an excellent base.

Answer More and more actuaries are getting involved in the investment side of the business, particularly with DCAT [Dynamic capital adequacy testing]. Although not everyone will use it, it is a good idea to be familiar with it in order to be a well-rounded actuary. Theory around cash flow and duration matching are also in common use. I believe this is now being covered in the CAS Course 8.

Answer In pension, you have the promises made to a participant to receive a pension, but you also have the employee and employer contribution that make up the assets. You need to know about investment.

Answer CAS only: finance is not used *per se* in our day-to-day work. However, knowledge about the effects of diversification may prove to be useful with respect to *planned growth* in P/C. This would relate to actuaries who have a more strategic role—at the executive level, or close to that level. Knowledge about the risks related to various investments (bonds, stocks, etc.) may prove to be useful in discussions at a higher level (executive level). Generally, knowledge about finance is very good to have, although the actual use of it is limited in the day-to-day work. No link with exams.

Answer Finance related to good consulting when valuating liabilities.

Answer Investment science.

Q Why do actuaries need economics? Please give examples and relate them to the SOA or CAS examinations.

Answer Great background to have for working in retirement or asset consulting so you understand more of what is going on *in the real world*. The only thing sometimes is that economics is a very theoretical science. Sometimes it is difficult to see a real-world application to some theories seen in Course 2, for example.

Answer To understand the link between the liabilities of a plan and the assets underlying the plan.

Answer A lot of our work depends on finance. For example, with the market situation today, pension funds are losing money. This fact should guide actuaries when they give advice to their clients on when to file an evaluation or the decision to improve the plan, for example. It is tested in the examinations in Courses 2 and 3.

Answer Economics is a big part of the second exam.

Answer Economics are also needed to understand the basics of actuarial mathematics. In Course 2, competency in economics is tested. This knowledge and skill will also be used in the workplace and serve to understand the ways a company and the market work.

Answer Actuaries need to be able to understand the structure and the workings of the different investment markets in order to manage their assets that back their liabilities well. Course 2 and Courses 5 through 8 touch on this.

Answer Set appropriate economic assumptions for actuarial valuation: discount rate, rate of return on assets, etc.

Answer The thing I remember about my economics class is the *marginal cost theory*, which I apply very often. But I'm not sure if I needed this class.

Answer Course 2. Also, the ideas of balance sheets are crucial even for pension plans and for the reserves of an insurance company. Pension actuaries must weigh the assets and liabilities of a pension plan against each other.

Answer Actuaries should have some idea of how macroeconomic events in the economy may affect the sectors of the economy that have an impact on their business. For example, how will a slowdown in inflation affect long-term interest rates?

Answer Actuaries need economics since almost all assumptions are based on current economic market conditions with projections for future economic outcomes.

Answer Some economics concepts can be used in modeling, for a better understanding of the impact of rate changes, for example. Simple concepts such as the law of supply and demand.

Answer CAS only: economics is not needed in our day-to-day work. Knowledge about it may certainly come in handy from time to time, but then again, more at a higher level (executive level).

Answer Depending on the field, it is not always necessary.

Answer Economic knowledge is needed to try to understand the needs of clients.

Q Why do actuaries need risk theory? Please give examples and relate them to the SOA or CAS examinations.

Answer Risk theory helps me understand the foundation of actuarial science. It is very important, I think, to be strong in this technical area since the ideas involved come up on a daily basis.

Answer Basic to our job is managing the risk related to a plan.

Answer I rarely use risk theory at my level. But it is important for the examinations.

Answer Course 3 tests these skills.

Answer Actuaries are trained to put a value on risk and handle future contingent events. Risk theory is the real fundamental bridge between life contingency theory and the business of insurance. Courses 3, 5, and 8.

Answer Understand risks, model risks to eventually put a value and cost on it.

Answer Risk theory is the heart of actuarial work. An actuary is an expert in the assessment and management of risk.

Answer Risk theory is the basic building block of the P/C business. However, as indicated earlier, the level of sophistication is rather lacking in the Canadian marketplace. However, it is helpful to understand risk theory in order to perform the daily work of a P/C actuary.

Answer CAS only: This is the basis of the pricing work in P/C. I cannot say, however, that what I learned in school with respect to risk theory helped me in my work.

Q Why do actuaries need loss modeling? Please give examples and relate them to the SOA or CAS examinations.

Answer I guess it is very important in CAS, but is less important in fields such as asset consulting.

Answer I am not yet familiar with loss modeling.

Answer I think this is more of a CAS thing or perhaps also a reinsurance thing. You need to be able to calculate the probabilities of incurring a loss before you can accurately set a price for an insurance premium. Loss modeling comes up in Course 4.

Answer CAS stuff. Used in pricing products by modeling future expected losses. Needed since non-life risks generally have the following characteristics: time of event unknown (so need a frequency distribution) and size of loss unknown (so you need a loss distribution).

Answer More useful for CAS, I think.

Answer Loss modeling is a fairly useful tool that is hardly ever used, at least in my experience. Lack of size (and therefore lack of data) is one of the problems encountered when trying to do loss modeling. Often a lack of time and resources will also force a company to use a broad-brush approach in its pricing and reserving modeling.

Answer CAS only: loss modeling may be used to forecast the severity of certain events, and also to determine how variable results will be from one even to the next (link with credibility of results). For example, in looking at automobile theft, vandalism and fire, loss modeling may be used to determine the shape of the curve that best describes severity (average cost). Once this is done, one can determine how variable this severity will be, and therefore how many observations are required in order to get credible estimates.

Q What stochastic ideas and techniques do actuaries use? Please give examples and relate them to the SOA or CAS examinations.

Answer I know areas of the actuarial field where it is extensively used and is important. This is not yet the case in asset consulting (at my level). But I know that stochastic ideas are very important in asset and liabilities management, an area I would love to get into later on in my career.

Answer The only method used frequently is the Monte Carlo simulation, mostly for the projection of the assets of the plan.

Answer Continuous Markov chains are used by actuaries and are tested in Course 3, I believe.

Answer Becoming more prevalent, especially with modeling possible future interest rate patterns when determining reserve amounts for life insurance and annuities. Also used for sensitivity testing and pricing of minimum guaranteed death benefits for segregated funds. Course 8 had a big section on this. Course 7 Pre-test had this.

Answer Projections of pension plan assets or surplus based on stochastic distribution of future interest rates. Can then determine the future distribution of values by percentile, calculate the probability of having a value less than some fixed amount, etc.

Answer To forecast what are best and worst case scenarios under different sets of hypotheses for surplus or deficit in a pension plan.

Answer Stochastic modeling of the cost of face amount guarantees on segregated funds.

Answer CAS only: stochastic techniques are not widely used in Canada. They may be used in the area of DCAT [dynamic capital adequacy testing], although I do not know anyone who has programmed or is using a stochastic model in Canada to do DCATs.

Answer More in asset consulting than liability consulting.

1.6 Supplementary Skills

In addition to being good in mathematics, economics, and other scientific subjects, actuaries need to have broad arsenal of other skills. What college and university courses should they choose to acquire these skills?

Q **Which are the most important complementary disciplines for an actuary and why?**

Answer Finance and accounting. The actuarial profession, especially in pensions consulting, is increasingly exposed to managing and considering the asset side of the balance sheet. A well-rounded professional therefore should have exposure and an understanding of accounting requirements and financial opportunities related to pension asset investments.

Answer Software skills (not really programming, but a good working knowledge of Microsoft Office is key). Interpersonal skills, communication skills (used every day in a working environment), understanding of financial markets and economics (it's my job!).

Answer Finance and economics. The pension industry is driven by the assets behind pension plans.

Answer I would say two: finance and computer science. It's essential to know about finance because everything that we do is related to finance. We use annuities, rates, present values, and so on. Understanding the value of money over time is essential. Also, because it would be too complicated to calculate everything by hand, we use computers a lot. I've seen someone with a Master's degree not getting a job because he couldn't work with Excel. Computers help us do our work faster.

Answer Statistics and probability, risk theory. Everything relates to this.

Answer For a pension actuary: I think accounting is becoming increasingly important in a consultant's job. Companies (especially the larger ones) are more concerned with the annual pension expense and it is key that actuaries have a good accounting background. A strong understanding of financial concepts is also very important (these courses are also useful for the later SOA Exams 6 and 8). Programming courses (Microsoft Visual Basic) are useful for beginning actuaries.

Answer For the first three actuarial exams, finance and economics are helpful. During my work in a casualty insurance company, I used a lot of programming skills (SAS and Microsoft Visual Basic). These skills are used to extract data from huge databases to do calculations. During my work in consulting (pensions), I used some accounting and language skills. In preparing reports, both skills were useful.

Answer Finance: knowledge of balance sheets and understanding the impact of our work in the real life!

Answer I would say that business courses are necessary to complement a good actuarial program. Courses such as finance, economics, management and marketing are essential for actuarial students. Communication courses, as well as language courses could be an asset. Computer science courses are also very important.

Answer Computer science—work with computers every day and it's not all programming. Investment and economics—much of an actuary's job is understanding how investment markets affect the risk assumed by an insurance company; medicine—used in underwriting business; law—understand contract law and legal regulations; communications—both written and verbal are important, especially when dealing with auditors; marketing—have to convince people to buy your products or services.

Answer Chartered financial analyst: specialist of the asset side of pension plans. Micro/macro economics: to determine appropriate economic assumptions. Communication: for presentations, understandability, and so on. Computer programming, etc.

Answer As an actuary just starting out, I find programming skills extremely valuable. Most junior actuaries will be required to program in a variety of languages. Having good software skills in general will always make for a more efficient actuarial analyst. Other disciplines that can be useful are economics and finance. As an actuary's career develops, softer skills such as management, delegation, verbal and written communication and relationship building can play an important role as well. However, actuarial students rarely consider them essential when they are still in school.

Answer Computer science early in your career because that is what you do. Then business or law should be helpful.

Answer Strong software skills are necessary. Microsoft Excel and PowerPoint. Programming is also necessary. These are all tools used in doing the work. Doing calculations and finding results require broad and technical under- standing of economics and finance. We have seen this year the negative impact the stock market downturn has had on the assets of a pension plan. It is important for an actuary to have a feel for the economy and how it is related to his work.

Answer Finance and economics: useful for seeing the *big picture* and broader context of the actuarial field. Ideas from finance and economics are required when making assumptions about future pricing of actuarial products. Com- puter science: you need to know how to program. Period.

Answer Actuaries have such a versatile training and have a lot of competencies that they can excel in a variety of work—finance/risk/insurance related.

Answer Liberal arts, because you can learn all of your actuarial skills on a company-supported program of self-study, but we will not pay a dime for philosophy, linguistics, or novels of the 19th century. A well-rounded actu- ary is more valuable to us in the long term than one who has had a narrow technical education.

Answer Accounting and finance for obvious reasons (as you go up in the hier- archy of an organization, it becomes really important). Management and human skills—too many actuaries lack these skills (and still become man- ager because of their professional status, e.g., Fellow).

Answer Economics: to project future economic scenarios/being familiar with current economical situations; finance and accounting.

Answer It would be finance, economics, management, and computer science. This view is based on my experience as a CAS corporate actuary. However, I believe the same would be true for SOA or for a pricing actuary.

Finance is important because when dealing with the accounting department. You must be able to speak their language. Proper understanding of financial statements is also important in a wide variety of situations (not just in your own company, but also, for example, for the pricing of some insurance products for corporate clients). Finance includes an understanding of investments, an area in which more and more actuaries are getting involved.

Economics is important to understand the concept of supply and demand in order to make good pricing decisions. If an actuary wants to climb the corporate ladder, then management skills are invaluable (not only in order to manage your employees, but also to meet your boss's expectations).

In addition, all actuaries have to be able to do some form of programming at one point in their career.

Answer Accounting and finance are very useful for valuation type work in the life insurance industry.

Answer Software skills, French and/or English (verbal and written), accounting, economics.

Answer Finance: to know about assets and liabilities. To understand a client's needs.

Answer Accounting: the knowledge of accounting concepts is crucial when using an insurance company's annual statement. Finance: financial theories are often applied to actuarial concepts and widely used in actuarial calculations (e.g., discounting of liabilities). It is also important for an actuary to understand the investment portfolio of an insurance company. Economics: principles of economics principles can be used to solve actuarial problems (e.g., analyzing supply and demand curves to price insurance premiums). Management: actuaries eventually get to a level in their career where they have to supervise and manage others.

Answer Communication skills cannot be overemphasized! Actuaries have always had a well-deserved reputation as "bright guys everyone wants to avoid." An actuary's ability to solve a problem is worthless unless he can persuade a non-actuarial audience that the solution substantially meets everyone's needs. Computer literacy is also important because almost all complex actuarial problems are solved with computers in today's environment.

Answer Computer science, because all of the more junior positions require programming skills.

Answer Investments and economics: the complement of liabilities is assets.

Software Skills

Here is what the survey respondents said about the importance of computer skills in general:

Q What software skills should actuaries have and why? **Please give examples.**

Answer All Microsoft Office tools (especially Microsoft Excel), databases (often firms have their own database system that is learned on the job), and logic. Knowledge of time management software is a must to effectively manage time in and out of the office!

Answer Skills are more related to problem-solving approach than real programming skills.

Answer Knowing all Microsoft tools such as Word, PowerPoint, Excel, etc., well. Being comfortable with searching for information on the Internet.

Answer Actuaries mostly use Microsoft Excel and should feel comfortable using it. Since we are playing with numbers all day long, any software that performs similar operations can be used.

Answer Programming skills are needed. Also, Microsoft Excel is a commonly used tool and the actuary should be very comfortable with using formulas and editing data. Sometimes Microsoft Access is used for data modification or verification.

Answer Strong Microsoft Excel skills are required, I think, in every company.

Answer Microsoft Excel, Access and Visual Basic programming.

Answer Microsoft Excel is a definite must. It is used in the day-to-day routine of an actuary. An actuary should also have good computer programming skills and be comfortable with the Internet. Like in many careers, the computer is one of the basic tools of the actuary.

Answer Spreadsheets: lots of work is done on spreadsheets these days, instead of programming. If you can build effective spreadsheets you will save a lot of time and look better in front of your supervisor, who will be able to check your work more easily.

Databases: they are about the concepts of fields, items, relations between tables, and allow you to work efficiently. Word processing and touch-typing: not key skills, but you will do a lot of typing over the career, so why not learn it? Macro writing, in Microsoft Visual Basic, for example. This can make your life a lot easier when it comes to repetitive tasks.

Programming: this skill is required for particular jobs such as obtaining data from mainframe systems. And probably the most important skill: knowing which tool to use in a given situation.

Answer Microsoft Office or equivalent: Excel (including macros), Microsoft Access, GGY's AXIS, some basic programming skills, some computer operating system skills, mainframe computing experience. Most spreadsheet work is done with Excel. In addition, Access is used for data queries to get results for subsets of data. AXIS is an actuarial pricing and valuation tool. Mainframe computing is required in big companies with large blocks of business. Many companies also still use in-house programs created in APL.

Answer I believe that Microsoft Visual Basic for developing Excel macros is must.

Answer Microsoft Excel and Word, and the ability to learn quickly.

Answer Excellent knowledge of Microsoft Excel (how to be efficient with it), Microsoft Access. Be good at programming languages—this differs from company to company. Some use SAS, Fortran, and Microsoft Visual Basic. It depends. If you have a good basis in one, your brain is already able to think in a programming environment.

Answer Familiarity with Microsoft Office (Excel, Word). This is the industry standard. All of this can be picked up on the job. As a new recruit you will look good if your ability quickly becomes a center of knowledge for the unit, so it pays to deepen your expertise.

Answer Since nowadays most work is done on computers, software skills are a must. Spreadsheets are extremely important. Word processing software and presentation software are also important. Most companies have their in-house programs and software with which one must become familiar.

Answer I think everyone should have software skills nowadays. Actuaries, more specifically, need Microsoft Excel and Access skills (or more generally spreadsheet and database skills). Actuaries typically work with a lot of data. Even if extracting data via a programming language, the quantity may still require some processing through a database to make the information more manageable for analysis. Databases, however, are not suited for analysis (at least, in my opinion). One does require in-depth spreadsheet knowledge to perform regression analysis, Monte Carlo simulation, Bayesian estimate, etc.

Answer Actuaries should have good Microsoft Office skills, especially Excel, Access and Word. They should have strong programming skills as well. Entry-level jobs, in particular, require good software skills.

- ▶ *E-mail.* Capacity to use e-mail (obvious, I guess).
- ▶ *Excel.* Capacity to use Excel (most of the calculations are done in Excel).
- ▶ *Word.* Word processing: to write memos/documents (speed of typing is important). To be an actuary, one must like computers because about 90% to 95% of the work hours are spent on a computer.

Answer Good in Microsoft Excel, Word, and Access.

- ▶ *Excel.* Advanced knowledge of Excel is required since most companies use Excel to put together actuarial analyses.
- ▶ *Access.* At least intermediate knowledge of Access is required. The actuary who is able to run complex queries will generate better data as a basis for actuarial analyses.
- ▶ *Word.* Basic knowledge of Word is required to convey results and findings of actuarial analyses.
- ▶ *PowerPoint.* Basic knowledge of PowerPoint is required to prepare presentations to management or clients.

Answer Capability to program because most junior positions require programming.

Answer Microsoft Excel: tables, charts, very powerful.

Programming Skills

Here is what the survey respondents said about the importance of programming skills in particular:

Q Which programming languages do actuaries need and why? Please give examples.

Answer Rarely. I need to write macros in Excel. That is about it.

Answer None. All companies have their own software now.

Answer Usually, each actuarial firm as its own program, so I don't think there is some particular programming languages needed. Of course, Microsoft Excel and Visual Basic are used a lot. I would say that an actuary should know at least one of the common programming languages (Fortran, C++, etc.). With that knowledge, it should be enough to adapt to others.

Answer In my day-to-day work, I use Microsoft Visual Basic (for macros) from time to time. Besides that, it is mostly company-specific programs. Therefore, more than knowing a single language inside out, I believe it is more important to have a strong understanding of programming methodology.

Answer I used SAS in a casualty insurance company and in government, and Microsoft Visual Basic in all of my internships. Although I took C++ courses at university, I have never used this programming language.

Answer I don't use any language, but the logic behind it is used for company-specialized software for actuarial valuations.

Answer Microsoft Visual Basic and Visual Basic for Applications are often used in the field. Knowing how to program macros and use them is often a great advantage. APL, although now more and more scarce, is also a programming language that has benefits since it is still used in some insurance companies.

Answer Experience with any programming language for *real* applications (not just basic applications) is very important. It is important not just to program, but also to go back and change programs, both yours and those written by others. Such experience promotes logical thinking, modules, testing in units, good documentation, patience, etc. Good languages to learn are those with structure (e.g., Microsoft Visual Basic, Fortran) and APL (because it is so different, and helps you think of matrices).

Answer Actuarial tasks are becoming less and less programming oriented, now that programs such as AXIS do many actuarial calculations. APL and Microsoft Visual Basic are probably mostly used. Maybe some SQL.

Answer Basic knowledge of programming languages such as Microsoft Visual Basic or Fortran is necessary to perform valuations in an efficient way.

Answer Although the programming languages that are used vary by company, some of the more common ones that I have encountered are APL, SAS, Focus and Microsoft Visual Basic.

Answer Programming was useful at the beginning of my career, but seldom used today. Logical thinking is required, and then you have to be able to coach a *real* programmer towards what you want. You don't do it yourself anymore.

Answer The ability to learn a programming language quickly. I have seen my languages at work.

Answer It depends on the company—a lot of companies have in house software. We use Fortran, Access, and AXIS. In P/C insurance, SAS is often used. A good knowledge of Microsoft Excel is always an asset.

Answer Very few. Familiarity with Microsoft Excel is useful. Some areas still use APL.

Answer APL, SAS, and Microsoft Visual Basic.

Answer I would say that programming skills are an asset but not a necessity. I personally hate programming and have avoided it since birth.

Answer The programming languages used by companies can vary. But actuaries, at the entry level especially, need programming skills in order to retrieve data for analysis. Very few companies have programmers involved in what amounts to data mining. Most IT [information technology] programmers are business programmers, worrying about transactions. Basically, they worry about the transaction of writing a new policy or paying a claim (financial information), but do not understand actuarial concepts such as earning premium or accident year data.

With the proliferation of databases, programming language will continue to flourish. Good Microsoft Visual Basic or SQL query skills will be required to extract and work with information. One does not want to repeat a series of manual command every month, writing a macro is much more efficient.

A programming language in common use is SAS. Some companies also still use APL, although this is diminishing. Finally, some understanding of basic programming concepts and languages (such as Cobol) can help, for example, when you are working with the IT [information technology] staff trying to debug a rating system.

Answer SAS (manipulation and analysis of large data sets, generating actuarial reports), APL (for reserves calculations), and Microsoft Visual Basic.

Answer Actuarial students should have good programming skills. However, I don't believe it is crucial that students know one language over the other. If students have good programming skills, they will be able to learn the language used by the company they are working for relatively quickly. I do believe that being able to use Microsoft Visual Basic in Excel is an asset.

Our company still uses Fortran, although knowing Fortran is not a requirement to get a job at our company. However, having good programming skills are.

Answer I think most actuaries can learn programming on the job. Knowing APL, C++, etc., is not mandatory—in some roles they're not even used! Now profit testing programs like TAS or MoSes are being used more.

CAS only: SAS—this is used by most insurance companies and consulting firms to extract and manipulate data. *(This is a must.)* APL—less used today than in the past, but still used by reserving actuaries. Microsoft Excel (including Visual Basic for Applications)—90% of work is done in Microsoft Excel, to the exception of data extraction (this is not a programming language, but I thought I should mention it anyway).

Answer Fortran and valuation programs.

Answer Microsoft Visual Basic and SAS.

Answer APL, Microsoft Visual Basic and Visual C++. Microsoft Excel and database knowledge.

Business Skills

It is often said that good business skills are essential to succeed in the actuarial world. What are business skills? Many companies now specialize in the teaching of business skills. Let us take a look at a typical repertoire of one of these companies. The company *Learn*$_2$ (*see* Reference 14, Appendix F), for example, offers business skills courses which include:

- ▶ Appraising people and performance
- ▶ Articulating a vision
- ▶ Coaching and counseling
- ▶ Communication skills
- ▶ Conflict resolution
- ▶ Counseling and disciplining
- ▶ Customer service
- ▶ Decision making
- ▶ Effective presentations
- ▶ Giving clear information
- ▶ Interviewing techniques
- ▶ Leadership situations
- ▶ Planning and scheduling work
- ▶ Planning your presentations

▶ Relationship strategies

▶ Training, coaching and delegating

▶ Time management and prioritizing

▶ Setting goals and standards

It is of course true that not all of these skills are needed by all actuaries all of the time. Here is what the survey respondents said about the importance of some of these skills:

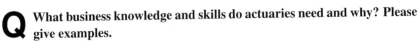

Q **What business knowledge and skills do actuaries need and why? Please give examples.**

Answer I am not sure I understand the question, but ultimately, selling skills and defending ideas can be great when meeting with clients.

Answer Understanding a client's business, including financial statements, calculations of profits, etc.

Answer At a higher level, actuaries sell services to clients. So actuaries need to be good in persuasion, understanding needs, and foresee problems or requests. Honesty is also very important. They need to be aware of the market in general. To understand their clients better, they also need to check specific fields in particular (if your client is a factory, you should know if the market is good for that field, not just for your client or in general). Actuaries also get to manage clients' teams: prices to charge, tasks to perform, who's to work together, time allocated to a project, and so on.

Answer A good background in business is necessary to an actuary. A knowledge of finance is essential in the study of actuarial mathematics, but even skills and knowledge in Marketing and Accounting can be useful since you will often find actuaries in the marketing and assets and liabilities management department of an insurance company. Since actuaries often hold management positions, management skills can be useful.

Answer There are others, but I would start with basic accounting (balance sheet, income statement, double-entry accounting) and finance (investments/assets characteristics).

Answer Knowledge of investment markets and types of investments, economics, insurance product structure, knowledge of how interest rates affect insurance liabilities and assets. Values of assets affect the amount of surplus that an insurance company has, which limits how much new business they can write. Values of liabilities affect reserves and surplus as well. There's plenty more, but that's all I can think of right now.

Answer Listening. Ability to solve problems.

Answer Understanding of accounting principles and financial statements, because you are sometimes responsible for big amounts on these reports.

Answer Ability to make decisions. This is often weak in actuaries, because they by nature see many sides to a discussion, and can accept many right answers. A bad decision taken is better than several good ones deferred.

Answer Organizational and time management skills are important—one must often juggle many projects at once with strict deadlines. Project management skills—knowing how to initiate, do, review and close a project with many variables and constraints, deadlines and objectives. Communication skills—verbal and written skills a must. Professional ethics—integrity, treating others with dignity and respect. Professionalism—a sense of professionalism in style, presentation and communication to others, verbally and in writing.

Answer This depends on the ambition of the person involved. Generally, the more ambitious, the more business knowledge and skills are required. Actuaries who are happy working in the back room and are not interacting with people other than their manager and co-workers probably don't need too many business skills. However, anyone who wants to climb the corporate ladder requires business skills. Indeed, let us not forget that this is what we are doing: running a business. The best actuaries in the field are, first and foremost, businessmen. They can understand the difference between an actuarial indication and the price the market will bear. They understand the implication on the company of their decision with regard to IBNR (incurred but not reported) loss reserves or reinsurance. They get involved in projects and understand the work flow of the organization, the difference between the bells and whistles, and necessary system enhancements. Business skills required include economics, marketing, management (both personal, time, and project), finance, investment, and communication.

Communication Skills

Here is what the survey respondents said about the importance of communication skills as their careers unfolded:

Q What communication skills do actuaries need and why? Please give examples.

Answer The more skilled actuaries are, the better they are, as I have found out since working full-time. Especially in Montreal, being able to speak and communicate fluently in both English and French is a *great* asset. For a junior consultant, it is of the utmost importance to communicate very well with the seniors so we understand exactly the work that needs to be done and once completed, to be able to explain it to the consultant in clear words. Listening is also a forgotten skill, but very important in day-to-day work. Presentation skills become more and more important, I assume, as you grow in the business and have to meet with clients and present them ideas and reports. Being able to support your ideas and organizing your thoughts are also key skills.

Answer Clarity, since it is difficult. Simplicity, since the client must understand.

Answer Knowing at least two languages, enough to be able to communicate, is essential. It is not unusual to encounter French-speaking clients, for example. Canada is so bilingual that it's not an option anymore. Also, an actuary needs to be able to express his thoughts and his knowledge. It will happen often that a more advanced actuary needs to explain something to a new one or even to a client. So being able to be clear, not too complicated and see when the other person doesn't understand is essential. The same skill applies for writing (the annual statements, for example, need to be clear, but simple).

Answer You need verbal skills to give presentations to your colleagues and to clients (particularly in the consulting field). Your writing skills will be useful to write actuarial evaluation reports or prepare internal status documents.

Answer Very good communication skills in order to gain credibility from people we are working with and to explain simply what we have done and why.

Answer Presentation skills are necessary since it is often required from actuaries to present their research results, projects or recommendations. Actuaries must also be able to *sell* an idea. In the consulting business, actuaries will interact with clients and need to be able convince the client of the necessity of a benefit plan for example. Actuaries also sometimes need to explain their results and recommendations. Hence communication skills are, as much as mathematics and business skills, essential in the making of a great actuary.

Answer Verbal: speaking to other actuaries in technical language, speaking to non-actuaries in non-technical language, presenting to management/board on reserves (appointed actuaries), presenting updated pricing models to underwriters, leaving phone messages. Written: documenting in clear and understandable language, writing some letters and reports (especially in consulting), e-mail. Listening: gathering information, learning about other areas of the company, learning other peoples' terms so you can speak to them in their language.

Answer Written: you often need to write important reports for management and regulators and/or auditors. Internal documentation of processes is needed as well. You also need to use e-mail effectively to communicate with non-actuarial staff, the field force, and customers who need to have technical concepts explained in non-technical language. Verbal: same reasons as above, without the written reports for regulators and auditors. Public speaking: you are often required to make presentations to audiences with varying actuarial knowledge.

Answer You need good writing and verbal skills. Mastering two or more languages is a must.

Answer The biggest challenge is to understand what you are doing and then to be able to explain it to people who don't have an actuarial background. Therefore, it requires excellent communication skills if you don't want to spend your life in front of your computer.

Answer Expression. It might often be hard to express a mathematical calculation in words, but this is a necessary ability. When working with a team, it is necessary to be able to discuss one's work and the need for certain calculations. It is important to be able to give oral presentations and to be able to speak in front of a crowd. In a corporation, you must speak in front of a group to share knowledge and ideas.

Answer You need to be able to communicate complex actuarial concepts to a variety of audiences and levels of understanding.

Answer Verbal and presentation skills to be able to present and *sell* your ideas and concepts to management. This is often critical when you work closely with upper management (such as corporate actuaries).

Answer Depending on the actuarial field, various levels of communication skills are required. For consulting actuaries, communication skills are extremely important. The level of knowledge of clients is quite broad; ranging from clients who are well informed to clients who have only a basic knowledge and depend on consultants to provide them with the required knowledge and information. A consulting actuary must therefore be able to communicate technical information into laymen terms and be able to tailor the information based on the level of knowledge of the clients.

Answer Actuaries need both written and oral communication skills, especially as more responsibilities are assigned to them. Actuarial mathematics is a difficult concept, and it is difficult to explain to a lay audience. Furthermore, at least with P/C companies, actuaries interact a lot with marketing, sales, and branch managers. In addition, all corporate actuaries interact with finance and upper management, as well as IT [information technology]. Rarely do actuaries price a product in a vacuum. The actuarial indication is only the beginning of the process. What good is it to price a product at the actuarially sound rate if nobody is going to buy it? Especially if the high price is driven from conservative assumptions. Pricing actuaries often have to explain or sell their recommended increases to a variety of people. When involved in various projects (or business), actuaries are often experts relied upon to help shape the requirements of the project. Good writing skills are essential in these instances. Corporate actuaries often need to explain their IBNR [incurred but not reported] loss reserve calculations to upper management. When a change in IBNR can erase the entire profit for a given year, not only are good communication skills necessary, political savvy is also essential! At a certain level, appointed actuaries also have to report to the Board of Directors. To avoid the "glassy eye" syndrome, good verbal communication skills are again essential.

Answer Actuaries can be faced with solving problems and giving answers to people within the company who are not necessarily mathematically inclined. Therefore, they need to be able to communicate and explain results in a simple way. Communication skills and the ability to explain technical terms and procedures are needed.

Answer (1) Capacity to explain complex concepts with simple words to non-actuaries (*extremely* important). (2) Listening skills are very important. (3) Speaking French is a big plus because of Quebec. (4) Written communication skills are very important (capacity to explain in writing complex concepts to non-actuaries, writing without mistakes). (5) Public speaking (presenting results to a group a people in such a way that the audience understands). (6) Being able to use non-verbal ways of communicating (visual aids for example).

Answer Written and oral skills are very important because there are a lot of reports to write and presentations to clients and employees.

Answer Communications skills are needed to interact with clients and other personnel.

1.7 │ Actuaries of the Future

One of the exciting aspects of an actuarial career is the continuously changing nature of the profession. As the world changes, so does role of actuaries in it. What will be some of the future skills required to adapt to this change? How do actuaries view the need for new skills as the profession evolves? In his introduction to actuarial modeling (*see* [10]), for example, Jones mentions several active fields of mathematical research such as neural networks, fuzzy logic, and chaos theory as sources for new actuarial techniques.

In some countries, more reliable statistical data, including mortality data, are now being actively collected. Over time, this new information will find its way into the actuarial world. New skills will be required to create credible predictive models based on these data. Knowledge management and global communications will continue to advance. International business will thrive and the need for foreign language skills will increase as a result. The globalization of industrial manufacture will create a need for new forms of insurance. The fusion of banking, insurance, and wealth management will create new actuarial challenges and opportunities. The list goes on. Equipped with appropriate future skills, actuaries will play a key role in making the risks inherent in these changes.

Here is how some of the respondents to the survey see the impact of these and other changes on the future of the actuarial profession:

Q **What changes in the knowledge, skills, and mathematical techniques in actuarial practice do you envisage in the next 5/10/20 years?**

Answer More software skills, more interpersonal and communication skills. More time management skills, since workload is getting heavier and heavier since machines can do work faster and faster. Fewer pure mathematics skills.

Answer More communication, more business/finance oriented.

Answer Forecasting techniques are becoming more widespread. Future actuaries will need to become more comfortable with liability projections and integrating future asset forecasts. Again I think accounting is becoming more and more important. The SOA examination in Course 8 has a greater focus on pension expense and balance liability than ever before. Also, more clients are becoming concerned with the effect that the pension plan is having on the company's books.

Answer I believe that in the future, the emphasis will be done more on the refining of the business skills and communication skills of a future actuary. More and more, the importance of these skills is growing and makes the difference between a good and a great actuary.

Answer The mathematics probably won't change, but the regulatory and computer knowledge will greatly change. There hasn't been a major new insurance product type introduced for about twenty to twenty-five years, so if one is introduced, then knowledge about how insurance products are priced and valued will change. Increased regulation and improved computer systems will allow risk calculation to become more fine and, hopefully, more accurate.

Answer Perhaps more Internet-oriented for basic information: services that are perceived as not adding value should be made available to clients on the Internet. We may be concentrating more on how to add value.

Answer More applications of fuzzy logic.

Answer CAS: In the next five to twenty years, actuaries will need to become more efficient as well as more refined in their approaches. However, the top priority will be to improve communication skills. *This is the single most important area for success!* Knowledge and skills: Actuaries will need to be more *non-actuary-friendly*. In the past, actuaries were thought of being in an ivory tower, but times have changed. More and more, actuaries are involved with a number of individuals in discussions leading to important business decisions. In the past, either the actuaries were not involved, or they would be one of the very few parties making the decision with the President of the company. This points towards the need for actuaries to be team players more than ever, while at the same time being able to influence people to make the right decisions. As a result, communication skills will be key to the success of actuaries in the future. Mathematical techniques: more specifically with respect to pricing in P/C, I foresee a much greater use of generalized linear models. With respect to dynamic capital adequacy testing, I would like to see a greater use of stochastic models as opposed to deterministic models. Will it happen? I do not know.

1.8 SOA and CAS

Q What is the main difference between working in SOA and CAS?

Answer SOA is a bit less technical and more regulations-driven than CAS. In SOA, you are dealing less with figures, and there is more people interaction (but I don't know CAS well enough to have a perfect answer to this question).

Answer International career opportunities, and the type of work itself.

Answer I believe the main difference is diversification. In SOA, actuaries tend to specialize more into one area, whereas in CAS, actuaries need to be more knowledgeable in several fields. There are advantages to both, and they fit two different types of people.

Answer On the CAS side, work is much more technical. You will have to calculate and to program a lot more. On the SOA side, you will have to work more with people outside of actuarial profession. You will have to do reports and to be in touch with clients.

Answer SOA: will generally deal with subjects more familiar to fresh out of school students. SOA deals more with life insurances and pension. CAS: is in itself more general and will require a more broad knowledge of the field because there are so many options out there. It is also more competitive and therefore more studies of the market will be done.

Answer As far as I know, the main difference has to do with whether you are interested about life related stuff (life insurance, annuities, investment products, pension plans, welfare benefits other than pensions, etc.) or non–life-related stuff (home insurance, car insurance, risk insurance, etc.). Also, it seems that consulting actuaries are put in the SOA group since most work on employee benefits–related issues, which are considered as life aspects.

Answer Although I have never worked in an SOA-related job, here is an extremely simplistic answer. Most CAS related jobs involve the pricing of automobile, property and liability insurance products. For example: personal car insurance, home insurance, small commercial businesses, liability coverage for doctors or lawyers, hole-in-one contests at golf tournaments, just to name a few. A main function of SOA actuaries is to price products that are more related to life insurance and pension plans.

Answer Good question, and difficult to answer since I've only worked in the CAS world (except two work terms, but one cannot get a good idea of SOA work from a few work terms). Off the top of my head, I would say that CAS work tends to be less technical than SOA. It seems to me that SOA has formula for everything (e.g., pricing life annuity), where most of the work is spent refining assumptions. On the CAS front, there is no accepted formula for pricing, just a basic methodology. Pricing work consists mostly of gathering data and projecting it into the future. Pricing in CAS is also very much dependent on competitive market forces. Again, the same is true of reserving, the setting of IBNR (incurred but not reported) loss reserves being mostly educated guesswork on the CAS side. On the environment side, CAS companies are generally smaller (especially in Canada), which would indicate greater opportunities for promotion. However, talented individuals will be able to climb the corporate ladder irrespective of their choice.

Answer I have never worked in P/C, so I would guess product line and the length of the liabilities from what I've heard.

Answer Only 9% of actuaries in Canada work in CAS. In my opinion, CAS work is less repetitive in the sense that you are more bound to be confronted to new problems than on the SOA side. This is the result of a few factors: The P/C industry is extremely competitive in Canada (insurers are always trying to find new ways to be more profitable and gain market share). P/C work deals with two random variables instead of just one on the SOA side. Pricing in P/C is based on 10 to 15 variables, which increases the complexity of the work of CAS actuaries. Because fewer actuaries work in CAS, smaller groups of CAS actuaries work in a given firm (in general, it may not be true in all cases), which leads to a *small-family* mentality. I am not sure about the SOA side, but the demand on the CAS side is currently increasing quickly and is bound to increase even more in the future. To the extent that this may not the case on the SOA, this could be another difference between the two career paths. Life insurers are profitable, P/C are not, or at least not as much. This is a major difference between SOA and CAS.

Answer Working in SOA (pension field) involves a lot of data manipulation at the entry level and for a few years afterward. In the CAS environment, actuarial analysts get more involved in the analyses (including data manipulation, of course) right from the beginning and get exposed to a better variety of projects.

Answer I can say from experience that the underlying actuarial principles are the same. The main difference is in the minutia of the regulations and how these have affected the practices and procedures of the industries.

Q When do you have to choose between SOA and CAS, and how easy is it to switch from one to the other?

Answer I always wanted to go in SOA since retirement and asset management are really the two fields I wanted to work in right from the start.

Answer Practically from the start. When you are too far in the exam process, it gets very difficult to go back and go to the other, especially when you are already working.

Answer No idea.

Answer The first four exams are jointly given by SOA and CAS. I guess you could always go back and switch track after the fourth one, but it would be easier to do so before. Also, once you have a job in one field, people usually stick with it. So I would say that the best would be to pick a track when you chose your job. You can also switch, but I don't believe that any credits would be given for more advanced exams in the other field. It could be discouraging to start over.

Answer You used to make the choice before Course 5, but the CAS society has decided to write their own Course 3, while Course 1, 2, and 4 are still the same for both SOA and CAS. Therefore, I believe students will need to choose much earlier, because they are realizing that it would be more appropriate to include more related-fields questions in the preliminary exams.

Answer At Concordia University, the syllabus does not require us to make a choice between SOA and CAS. Thus as long as you have done the first four exams or as long as you are not looking for a permanent job, you don't have to choose. However, since the CAS Course 3 will in future be different from the SOA course, students may need to make a decision sooner than I had to. For a student, the only difficulty in switching from one side to another is the exams. When you are working it might be harder since you cannot get experience in both field at the same time.

Answer You have to choose between SOA and CAS after having written Course 4 and it is possible to switch from one to another, however this requires to start over the Course 5 and up in order to get an FSA or FCAS.

Answer Can be done at any stage. Is easiest if you have no more than the first four exams (so you don't have to rewrite exams) and no more than three years of experience. This way your compensation will not fall significantly when switching. I know many people who switched from SOA to CAS while working full-time. When I switched, I had five CAS exams, approximately six SOA exams (old system) and less than one year full-time experience. I took a 10% pay cut. We have another analyst who switched after two exams and two years experience. I don't think she took a pay cut. In this case, she was able to transfer to our company from our affiliated life company. It is now less common than it was for companies to have life and non-life divisions (e.g., CGU Canada, Royal, and SunAlliance), which can offer the best opportunity to switch.

Answer The first four exams are jointly sponsored by the SOA and the CAS. So switching isn't too hard. However, the work can be quite different from what I've been told, so it may be more difficult to do.

Answer Should probably choose before starting exams, but ultimately you have until Course 4 to decide since Courses 1 through 4 are jointly administered by SOA and CAS. After that, I think it is just a waste of time if you switch since exams passed in SOA or CAS will not be credited in the other organization.

Answer Since there are so many exams to write, you do not want to switch too late!!! Exams will be different quite soon.

Answer Most students do not have to choose between the SOA and the CAS until they graduate from university and are looking for their first full-time job. The first four exams are common to both the SOA and the CAS (this will change in 2003). It is only with Exam 5 that a choice has to be made. Since extremely few students graduate with five exams the decision as to which Part 5 to write will have already been made based on their choice of jobs. I do not know of many actuaries who have made the switch, and the few that I do know, made the transition prior to having completed four exams.

Answer Too early. I felt I was "tagged" SOA from the beginning, and only those who chose to go a "different" way went to the CAS.

Answer First four exams are offered jointly by the CAS and SOA. They become separate afterwards. From one to the other, I guess it depends how willing an employer is to be to hire someone with a certain number of exams and no experience in the field at all since I believe CAS and SOA jobs don't relate that much.

Answer I would think shortly after you start your working career. You are unlikely to be able to switch within the same company.

Answer In my case, I had to choose because of my professional exams. I believe that it is not that easy to switch from one to another. Maybe the group insurance as well as the finance side are the best "points of contact."

Answer Currently, the first four exams are administered jointly between SOA and CAS. The remaining exams are to be written either through SOA or CAS. I have no experience with the CAS exams, and no comment on how easy it is to switch between associations.

Answer Ideally, you should choose before writing any SOA or CAS specific exams. If no specific exams are written at the university level, than you should choose before starting to look for a job. Universities with internship programs are quite useful in this regard, providing students with opportunities to work in different types of environment. If, after working in a specific area (say SOA), you decide that you would rather work in another area (CAS for argument's sake), it becomes difficult to switch. For one, any new employer is unlikely to give much weight to the experience in the other field. Depending on the length of time spent in that field, this may mean a sharp decrease in salary for switching. Also, if any SOA specific exams were successfully written, this investment is now wasted. Finally, any new potential company will wonder why you are switching. There may be some worry that you will switch back, since the grass is not always greener on the other side. Although not impossible, the switch becomes increasingly difficult and costly the longer you wait.

Answer Since Courses 1 through 4 are common to both societies, one can choose when ready to write Course 5. Ultimately, experience gathered in any of the two fields is valuable experience. It can be useful if later a change is made.

Answer You have to choose between SOA and CAS after having completed Course 4. After Course 4, exams are track-specific. It is not an easy process to switch from one track to the other, unless one is willing to write additional exams (there is no credit given for courses beyond Course 4 from one track to the other). However, if one works in a consulting firm where there is an SOA-related department and a CAS department, one might have a chance to work in both departments. Although I am pursuing the SOA track, I have had the chance to work in P/C as well. My company has both a life and P/C department.

Answer You should probably choose after you complete the first four exams. It would be fairly easy to switch on or before that time. After that, you are starting to get more specific and it makes the switch more difficult. I made the choice after completing college. I opted for SOA since I had followed this path in college and had taken more courses in this area.

Answer The first four exams are common, so this gives the student a chance to think about it. The best thing to do is to try both (if possible)—an internship program is perfect for this. The earlier it is in your career, the easier it is to switch. At the beginning of your career, it is extremely easy (this is what I did).

Answer Very early in the process. Don't think it is easy, it is very different.

Answer I think the decision should be made before graduating from university. Even though the first four actuarial exams are jointly sponsored, the experience acquired is totally different when working in SOA versus CAS. Also, employers are generally reluctant to hire someone who wants to switch. In some insurance companies or consulting firms who hire both types of actuaries, a switch might be easier than going on the market looking for another job. I strongly recommend students obtain at least one internship in each field so they can be aware of the differences and similarities between SOA and CAS.

Answer The SOA and CAS cosponsor the first few exams. The later exams
are different because the regulations of industries are different and this has
had an impact on practices and procedures. Typically, one chooses a path
after one completes the cosponsored exams but that doesn't always happen.
Some actuaries choose to complete both and others choose a track before
they begin. Switching is not difficult as long as one understands that his/her
body of knowledge accumulated to date will not be more than a rough guide
in the other industry.

Answer You should choose before you start writing the exams.

1.9   Actuarial Accreditation

Q How important is the number of actuarial examinations passed for an
actuarial career? Illustrate your answer with examples of careers in
companies you have worked for.

Answer It is of a great help, but in my company we have *excellent* people who
are successful consultants and are not (and will never be) Fellows of the
SOA. Of course this is rare, but it shows that it can happen. Personally, I
hope I can make it to the end of the exams system and become a Fellow.
But I sincerely think that I wouldn't be any worse off as a consultant in
asset management if I were not a Fellow. I know people who are Fellows,
but are not very good consultants because they are missing other essential
skills.

Answer Exams bring recognition. Only the title is important. Not having the
title makes the things a bit more difficult, but not impossible.

Answer If you want to advance in a company, you need to be at least an Associate. I see many successful actuaries who are in charge of many clients
and have stopped writing exams after their Associateship. But it takes them
more time than it does a Fellow to get to the same point. It depends a lot
on what your goal is and what other skills you have. A Fellow without
interpersonal skills would probably lose clients.

Answer Consultants emphasize a lot more the importance of exams, since you
are not allowed to sign anything for any clients until you become a Fellow.
In insurance, I have met many people who chose to stop writing exams,
and who have great jobs, and great positions in the company. Therefore, it
depends on the goals of the student.

Answer I think it is directly correlated with how fast you can move up in a
company such as a pension consulting firm. Showing dedication towards

passing exams combined with gaining experience on the job can make for faster promotions.

Answer Usually, Fellow actuaries have jobs that carry more responsibility. For example, in an insurance company, they are often head of a department. As such, they coordinate the work of a team. In a consulting office, they are senior consultants. This means that they manage projects, meet clients and do the reports. When you are not a senior consultant, you are the one who does the evaluations, a more repetitive task.

Answer Not as important in group insurance benefits since most of our work does not require the signature of a Fellow.

Answer I do not believe that it is absolutely necessary to finish the actuarial examinations. In some of the companies where I have worked, I have met many actuarial professionals who stopped at Courses 4, 5, or 6 for reasons of their own. This has not stopped them from getting promotions or acquiring more knowledge. However, one cannot say that completing one's actuarial exams has no effect. From my experience and from what I have seen, successfully completing exams fast-forwarded more than one person's advancement.

Answer Canada: in life insurance company or pension consulting: can't move past analyst level without finishing exams. In casualty consulting: ACAS (or close) with experience can be a consultant, but can't sign reserves. In small/large casualty insurance companies: can't move into actuarial management, but may be able to move into non-actuarial management, e.g., underwriting.

Answer You need to get them all to become an actuary. The industry demands that. Careers can be very valuable to a company—in some cases they do the same work as actuaries, but are not allowed to officially sign anything. I feel that work experience is more valuable than exams. In the United States, individuals who have their ASA can call themselves *Actuaries*. There are many state commissioners who only have ASAs, but have a wealth of experience and are just as capable as their FSA counterparts. But companies don't hire people to become Career ASAs—they expect them to finish their exams.

Answer No exams: must concentrate on other skills such as communication, marketing, etc. Up to four exams: proven mathematics background ASA: May be sufficient for many actuaries. FSA: allows you to use the designation fully; opens all the doors.

Answer I chose to stop at the ASA level and I'm really happy about it. I'm pursuing a career as a human resource specialist in employee benefits, and although I do not have the title to sign any official paper, I feel that I have enough experience and knowledge to smartly challenge our consultants, and not just sit back and listen to what they have to say.

Answer There is a big gap between an FSA and a Career ASA. If you are progressing steadily toward FSA, then any difference in pay, related to number of exams, will even out by the time you are fully qualified.

Answer It really depends on the organizations that you work for. As mentioned in a previous question, in my current organization, there are many actuaries that have not finished all actuarial exams that have excellent positions in other departments such as claims, underwriting, information technology, and finance (all Senior Vice-Presidents or Vice-Presidents). Also, within the actuarial department, there is one business unit that is run by actuaries without many exams. In other companies, you wouldn't see that.

Answer In consulting, examinations are very important if one wants to be a consulting actuary entitled to sign valuations. In large companies, it is harder to "move up the ladder" if you aren't a fully qualified actuary. It seems that in the past, ASAs were considered for senior positions. Based on what I am seeing in my company nowadays, that isn't the case. Qualifying is very important. Colleagues that choose to not complete exams are eventually placed in more technical positions, with lesser responsibility.

Answer At the end of the day, the number of actuarial examinations passed shouldn't matter, since one can learn the same thing on the job. However, in most cases, it does matter. One prior company I have worked with rewarded good work rather than exams. There, only two of the six managers were Fellows, the others having stopped writing exams. In most other places, having a Fellowship will tend to open doors. In some instances, Fellowship is an absolute requirement because of regulations (signing of valuation report, rate filing with regulatory bodies). In Canada, where the Associateship is not recognized, you almost need your Fellowship to become a consultant. Generally, Fellows will tend to move up the corporate ladder faster. Fellowship tends to indicate dedication, creative thinking, hard working qualities. However, there are always exceptions, and a Fellowship is generally not an indication of good management skills (besides time management). Companies would be well advised to look outside the box for promotion. All this being said, I do believe that passing examinations is still very important for an actuarial career.

Answer The number of examinations passed usually increase an actuary's responsibilities in the every day duties. More examinations represent more knowledge as well as commitment to the profession and to one's career. An example is someone whose title changed to Director of Actuarial Services when Associate Status, as well as sufficient experience, were reached. The new position encompasses managing the projects of a team of actuaries and reporting to the Actuarial Vice-President.

Answer I believe that the number of actuarial examinations passed is important. The degree of importance varies from one company to the other. A person who has a good success rate with the exams is viewed as a very serious person. All the actuaries appreciate the level of difficulty, discipline and hard work necessary to pass those exams. Therefore, a student who consistently writes them and passes them is viewed as a disciplined person and a hard worker. Depending on the type of work, not becoming a Fellow can stop one's progress within an organization. I work for a consulting firm. Some clients specifically ask for a Fellow to perform certain assignments. I work in the life valuation area. The appointed actuary needs to be a Fellow. Therefore, the fact of not becoming a Fellow has a great impact on one's career path because this person will never be able to become an appointed actuary. Moreover, at our firm, certain job levels can only be reached by Fellows. I have given examples on how being a Fellow is important. However, there are many companies where this is not as big an issue. I know a lot of actuarial people who did not become Fellows but still have a great career. Not every job requires a Fellow. However, on a personal level, I do believe that becoming a Fellow is important. I chose to study in actuarial mathematics so I could become an actuary. Never becoming an actuary would have meant not completely reaching my goal.

Answer CAS only: compared to missing one exam with having passed all exams, the difference is like night and day. Here are a few reasons on having passed all exams):

- ▶ *Promotion.* More likely to be promoted.
- ▶ *Opportunities.* Numerous new opportunities (both internal and external). Generally, some exams, such as pricing and reserving, are key to the work done in P/C. Therefore, having passed them (or one of them)—although nothing is automatic—may lead to more responsibilities, a promotion, getting staff, etc.
- ▶ *Salary.* Usually, the salary and the title are a function of experience and the number of exams passed. Companies usually apply varying weights to the two components. However, one thing is for sure, more exams usually means a faster progression through the ranks.

▶ *Pitfall.* There is one pitfall to having passed a lot of exams, however. If you have no or very little experience, and have passed a lot of exams, your "employment cost" may be too high. Therefore, you may end up being offered a lower salary than what you should really get, or some employers may decide not to offer you a junior position because it would be too costly considering your lack of experience. I consider that four exams—up to a maximum of five—is not too many for someone without any experience.

1.10 │ From Associate to Fellow

Q **What can a Fellow do in your company that an Associate is not able to do?**

Answer Not much in asset consulting, maybe some asset/liabilities studies.

Answer Signs reports. Is client manager. Grows faster in the company.

Answer Not a lot, if the Associate has good complementary skills.

Answer A CAS Fellow can sign the actuarial reserves and the financial projections that all registered companies are required by law. An Associate will know, and will perform those calculations, but will not be able to sign.

Answer Besides signing valuation reports, I'm not sure.

Answer We are required to be a Fellow to sign actuarial valuations.

Answer SOA and CAS Fellows have more credibility and are given more responsibilities than Associates. Therefore, Fellows will often be given the final say in decision-making situations and are asked to give their judgment on a particular project. They are more trusted. As if they could *do* more than Associates. However, I believe that the *doing* is strictly correlated with the number of years of experience rather than the number of exams passed.

Answer Sign reserves and rate filings—manage other Associates or Fellows. Otherwise, no difference in the work or opportunities.

Answer A lot—an ASA can't provide an official opinion on anything. Also, ASAs are thought of as FSAs in training. So they aren't given the same responsibilities that an FSA has. Career ASAs are often put into other (non-actuarial) roles in the company.

Answer Sign actuarial valuation reports!

Answer In Canada, an Associate of the Casualty Actuarial Society does not have any more legal authority or signing power than somebody with only a few exams. It is only once actuaries become Fellows (FCAS) and then receive their Fellowship of the Canadian Institute of Actuaries (FCIA) that they can start to legally sign documents. Examples are: Ontario automobile rate filings, year-end reserve valuations, and DCATs [dynamic capital adequacy testing].

Answer My department could not afford an FSA!

Answer Sign the actuarial valuations of a pension plan! This is the main job of an actuary!

Answer Sign business unit valuation reports. Be Appointed Actuary for a subsidiary.

Answer I guess only what they can't do because of legal constraints (e.g., sign an Ontario rate filing or the Appointed Actuary's report).

Answer FSAs can sign off on actuarial valuations as a main signature, ASAs can only cosign on certain projects. FSAs appear to be given more senior roles and greater responsibility.

Answer In my former company, the only thing an Associate could not do was sign the year-end reports or rate filings filed with regulators. This is only because regulations require Fellows to do that. But there is nothing that says they cannot do the actual work, with the work being reviewed by the Fellow who will sign the documents.

Answer Fellows can sign actuarial valuations. They can be Appointed Actuaries. Some clients specifically ask for Fellows. Therefore, these assignments are not available to non-Fellows. The higher job levels can only be reached by Fellows.

Answer CAS only. Tangibly: sign actuary's reports, sign rate-filings for the Financial Services Commission of Ontario, call themselves *Actuaries*, act as proctors for actuarial exams. Intangibly: receive greater trust from non-actuaries because of "perceived superiority." In actual fact, it makes no difference.

Answer Sign reports.

1.11 | Going for a Master's

In countries where the education of actuaries is university-based, it is often customary to pursue graduate studies at the Master's level. Having a Master's degree is in some sense equivalent to having achieved Fellowship status in a professional society. A Master's degree is required, for example, to become an Appointed Actuary in a country like Denmark. Moreover, university diplomas in many countries of continental Europe such as Germany are equivalent to Master's degrees in the United Kingdom and North America. However, some countries are beginning to introduce Bachelor's programs in actuarial science. In Austria, it is now possible to obtain a Bachelor's degree in actuarial science.

In countries where the accreditation of actuaries is based on professionally set examinations, the need for higher degrees is not obvious. In Canada and the United States, actuaries have to spend many years studying for their professional examinations. As a result, there is little incentive for acquiring a higher degree after that. Here is what some of respondents to the survey had to say:

 Discuss the value of graduate studies in actuarial science, accountancy, finance, economics, MBA, and so on, for an actuarial career.

Answer My actuarial background has enabled me to leverage my MBA degree into a very valuable career as a Management Consultant. My post-MBA career, however, is not related to the actuarial discipline.

Answer I don't believe that it is of great importance. Maybe an MBA is helpful, but in my case, writing CFA [chartered financial analyst] exams was of greater value for a career.

Answer Very limited value. Experience makes a good consultant, not study.

Answer Studying in actuarial science helps a lot for the exams. I also believe that it is easier to start. We have enough material to learn when we start, if we at least know all the actuarial science material, I believe the progression will be faster. But it is important to take courses in other fields to complete our knowledge.

Answer I do not know anyone with a graduate degree.

Answer I think that the SOA and CAS exams can be much more helpful in an actuarial career than any graduate studies. The only exception may be for actuaries who work in the finance field.

Answer In the actuarial field, SOA and CAS examinations have a greater value than graduate studies in actuarial science or other. However, they are not overlooked, and I consider it as a sign of a person's interest in learning and furthering their education, a fact that can only be applauded. A deeper understanding of actuarial science, finance, economics, etc., cannot hurt a future actuary but is not as essential as the SOA and CAS examinations.

Answer Good if planning a more research oriented career. I think there are many needs to fill.

Answer For me, the most useful tool would be project management. Other than that, being a recognized actuary (even ASA) is enough, most of the time, to add serious weight to the advice you give.

Answer I don't know yet. I don't think that I will pursue a Master's degree in actuarial science, I am yet to encounter a person in the workforce who has done so. I have not closed the door to the other degrees mentioned.

Answer Very little in business.

Answer In my opinion, graduate studies have very little value in promoting a successful actuarial career. It is not worth the time.

1.12 Alternative Careers

What if you have spent many hours studying to become an actuary and that at some point you simply say to yourself that it is time for a change? Is there anything else you can do with all of this specialized training and knowledge? Here are some alternative career options:

Q **What are the alternative professional options for actuaries who decide not to pursue an actuarial career? Please give examples.**

Answer Management consulting (strategy or financial consulting with any of the major consulting firms, requires adaptability, ability to work well in teams, demands strong stamina to work long hours while traveling sometimes extensively), investment banking (demands strong interest in finance and ability to work long hours under pressure), asset securitization, risk management. An actuarial background and training provides a tremendously strong springboard to opportunities in a variety of non-actuarial disciplines.

Answer Investment managers, client relations for banks, managers, consulting firms, teacher, stockbroker, team leader.

Answer Mathematics teacher, investment consultant.

Answer Investment banker, portfolio manager, accountant, statistician, professor, researcher, economist.

Answer There are many fields available: teaching, research, finance, programming (for actuarial companies).

Answer I think an actuarial degree keeps many doors open as it shows an ability to understand complex concepts and apply theory to practical business problems. I know actuaries who have become directors of human resources for large companies. This position would involve the company's compensation practice. Also, some actuaries decide to enter the finance field.

Answer Finance (asset management), human resource advisor, teacher.

Answer Graduating with a BSc or BA in actuarial science is a great basis to later on continue to an MBA. Any profession in finance and management can be pursued. As well, becoming a CFA [chartered financial analyst] or chartered accountant are possibilities. Of course, the option of working for an insurance company is always there.

Answer There are many professions that use the mathematical skills, technical skills, software skills, and problem-solving skills that are supposed to be developed in the actuarial education process. Certified financial analysts, statistical modeling specialists, forecasting modeling specialists, computer programmers, etc. Generally in investment or IS-type employment. Could also sell insurance with an actuarial background.

Answer Financial analyst with investment managers (requires the CFA [chartered financial analyst] designation), director of employee benefits with private firm, IT [information technology] specialist (requires strong software skills), banking specialist.

Answer Banking, finance.

Answer Interesting question. One would be to pursue the chartered financial analyst designation. There are three exams and you can work in more investment related fields and skip the mathematics portion. Another would be to go to graduate school in order to teach or to get a degree in financial engineering. The field is opening a lot to non-traditional positions. I'm not that familiar with them.

Answer Produce special quotes: be more systems-oriented (if you enjoy programming), become a manager in an insurance company where business skills are very important, doubled with actuarial knowledge.

Answer Teaching, risk management (derivatives, hedging programs, portfolio management), teaching mathematics at the high school level.

Answer I believe that there are many opportunities. Within my current organization, there are many actuaries outside the actuarial department: claims, underwriting, investment, and IT [information technology] (all vice-presidents). Outside the insurance world, you could probably become a financial advisor, a management consultant, or a risk management consultant.

Answer Here is a list of professions some of my friends chose after deciding not pursue an actuarial career: banking, investment banking, financial advising, brokerage services, pension administration, programming specialists, investment consulting.

Answer Teacher, statistician, finance-related profession, programming, etc.

Answer Engineering, consulting, finance/investments, banking, statistician, math teacher/professor.

Answer Programmer, teacher, career in research, CFA [chartered financial analyst]. There are definitely many career possibilities for someone who has passed actuarial exams (especially is all of the exams have been passed).

Answer Teaching, working in finance.

Answer Retraining would be necessary if an actuary expects to compensated as highly, but there is no limit. There are a number of ancillary positions within the insurance/pension industry to which actuaries could easily apply their skills. Sales/marketing, software development, accounting, business planning. They could also readily move into another part of financial services and become stock analysts or financial engineers.

Answer Computer programming, marketing.

Answer Teaching at the college and university level. Investment: I know a few people, they wrote their CFA [chartered financial analyst] exams and now work in marketing or servicing departments of investment management firms.

1.13 Actuaries Around the World

As is the case with most formally structured professions such as medicine, engineering, accounting, and others, the international employment of actuaries involves two critical elements: a recognition of academic qualifications attained in another country, and a license to practice.

In Europe and Latin America, actuaries have tended to qualify by completing a course of actuarial study, usually up to the Master's level, at universities accredited by actuarial societies or governments, and by meeting certain professional requirements.

In Great Britain and Commonwealth countries, the Faculty of Actuaries of Scotland and the Institute of Actuaries of England have defined an actuarial syllabus and sets of examinations based on this syllabus. Students in many parts of the world take these examinations to become actuaries in their countries. Certain university courses at designated universities can be credited towards this process.

In the United States, the Society of Actuaries and the Casualty Actuaries Society have defined a syllabus and sets of examinations that must be taken to become an actuary. No university program or courses are credited toward these examinations. Most American and Canadian actuaries and many others around the world become actuaries by passing these examinations.

As a result of changes in the world economy, several countries with a university-based accreditation system are looking toward instituting nationally administered examination systems, whereas countries with professionally based accreditation systems are considering the idea of granting exemptions for some courses taken at university. The dynamics involved in these deliberations are outlined in greater detail in Brown's paper on the globalization of actuarial education (*see* [6]).

The process of becoming an actuary is far from uniform. When contrasting the North American actuarial education with that of Europe, for example, one working actuary explains that "in general, the European programs are more like graduate programs or short seminar-and-test programs than the lengthy study-while-you-work exam system of the United States. As a result, the actuarial training tends to be more theoretical than hands-on, and the resulting actuaries are very strong in statistics, modeling, etc. American actuaries tend to have more real-life experience. Some of this focus may be due to the available data (at least from a P/C perspective). Regulatory requirements force insurers in the United States to collect much detailed data, (usually) suitable for analysis. European insurance data is not as voluminous, so there is more emphasis on theory, modeling, and the like. Purely anecdotally, the French actuarial education may be the most theoretical, while the UK exam system is the one that most resembles that of the United States. However, the United Kingdom does not make the distinction between Life and P/C (Non-Life, as they call it here) made in the United States (SOA versus CAS), and I believe many of the European actuarial societies also take that approach."

Nevertheless, it is actually fairly easy for actuaries educated in North America and the United Kingdom to work in other countries. The following examples will give you some idea of different national accreditation systems and of the portability of the acquired qualifications between certain countries.

Argentina

The education of actuaries in Argentina is university-based. It is necessary to obtain a degree of *Actuario* at a local recognized university in Argentina, like the University of Buenos Aires, which follows the syllabus recommended by the International Actuarial Society. (*See* Chapter 2.)

Nowadays the University of Buenos Aires issues diplomas following two orientations: *Actuario-Administracin* and *Actuario-Economa*, each of which provides an actuary with a license to practice. The degrees differ only in some courses on administration and economics.

To become accredited actuaries, graduates must register their diplomas with the *Consejo Profesional de Ciencias Econmicas* of the State where he or she would like to practice as independent consultants or be employed in a position that requires an actuarial degree, such as the *Consejo Profesional de Ciencias Econmicas de la Ciudad Autnoma de Buenos Aires*, for example.

Each State in Argentina has a public professional council called the *Consejo Profesional de Ciencias Econmicas*, created by law, which controls the independent activity of accountants, actuaries, administrators, and economists. These councils are responsible for maintaining professional standards. In order to be able to issue independent reports and formal advice, actuaries must usually be registered with the Consejo of the State in which they work. Registration requires an actuarial diploma issued by a recognized Argentine university (private or public) or a diploma from a foreign university recognized by a public university with a full actuarial program, such as the University of Buenos Aires.

Australia

Admission as a Fellow of the Institute of Actuaries of Australia (FIAA) is granted once all five parts of the Institute of Actuaries of Australia's (IAAust) education program are successfully completed: (1) Part I—Technical Subjects. (2) Part II—The Actuarial Control Cycle. (3) Part III—Specialist Subjects. (4) The Practical Experience Requirement. (5) Professionalism Course.

Part I is made up of nine subjects including statistical modeling, financial mathematics, stochastic modeling, survival models, actuarial mathematics, economics, finance and financial reporting and financial economics. All nine subjects must be completed.

Accredited undergraduate actuarial programs and non-award courses are offered by Macquarie University, Sydney, the University of Melbourne, the Australian National University (ANU) in Canberra, and the University of New South Wales (UNSW) in Sydney. Alternatively, these subjects can be studied by correspondence through the Institute of Actuaries (London).

Part II of actuarial education is the actuarial control cycle, which is an innovative means for learning how to apply actuarial skills to business situations across a wide range of traditional and non-traditional practice areas. Developed by the IAAust, this course is taught by four universities in Australia (as mentioned above). A strong and rigorous policy framework for accreditation of the university courses is in place, so that the IAAust maintains quality control of the teaching and assessment of the courses. After completing Parts I and II, members achieve Associateship of the IAAust (AIAA).

Part III consists of specialist subjects, of which students must complete two, in life insurance, general insurance, superannuation and planned savings, finance, and investment management. These yearlong courses are developed and managed by the IAAust and are offered by distance education.

Students must complete 45 full-time working weeks of relevant work experience after having completed Part II. Activities that qualify as relevant experience would include work that makes use of economic, financial and statistical principles to solve practical problems; work that deals with the financial implications of uncertain events.

The Professionalism Course is a highly participative three-day residential course conducted by the IAAust. It aims to facilitate knowledge of the obligations, risks and the legal responsibilities of being a member of the actuarial profession.

The IAAust has concluded a number of bilateral agreements for mutual recognition of Fellows with the Faculty and Institute of Actuaries (UK), the Society of Actuaries, the Canadian Institute of Actuaries, and the Society of Actuaries of Ireland.

These agreements enable actuaries to practice professionally in other territories subject to meeting the requirements of the local actuarial association. Each agreement is predicated on equivalent educational and professional conduct standards. In addition, a period of professional practice and residency within Australia is required prior to overseas actuaries being eligible to attain full Fellowship status of the IAAust.

Associateship is obtained by passing Parts I and II of the examinations. Moreover, "in order to enter the actuarial profession, graduates from an Australian or New Zealand university must have degrees with mathematics as a major subject, or at an Honors level in a non-mathematical subject, provided that a sufficiently high standard of mathematics has been demonstrated during the university course or at school." The Australian National University is accredited by the Institute of Actuaries of Australia to provide students with exemptions from certain examinations of the Institute, provided the students obtain sufficiently high grades in designated courses.

Austria

The Austrian equivalent of a Fellow of the Society of Actuaries is that of an *Anerkannter Aktuar*, a regular member of the Actuarial Association of Austria. To become a member of the Society, a candidate must have obtained a university degree in actuarial science and have three years of professional actuarial experience. The Technical University of Vienna is the only Austrian university offering novel degree programs in actuarial science following the North American Bachelor's and Master's degree structure. Austria is a member of the *Groupe Consultatif* and Austrian actuaries can take advantage of the European reciprocity agreements coordinated by the Groupe. Foreign-trained actuaries can become members of the Association if their academic training and professional experience meets the requirement of the Association. The main function of the Austrian Association of Actuaries is to promote the education and training of its members, to represent actuaries both nationally and internationally, and to establish guidelines and rules for good actuarial practice in Austria.

Belgium

The Belgian equivalent of a Fellow of the Society of Actuaries is that of a full member of the Belgian Society of Actuaries (KVBA-ARAB). How do you become such a member? Candidates must first obtain a Bachelor's and a Master's degree in mathematics, economics, civil engineering, or physics (which takes four to five years at university). They can then be admitted to an actuarial program recognized by the professional organization (run by ULB in Brussels, UCL in Louvain-la-Neuve, KULeuven in Leuven or VUB in Brussels). The students need at least two years to complete the program. On the basis of this academic training they can become junior members of the Belgian Society of Actuaries. They then need three more years to become full members. During that time they must follow certain courses on professionalism, code of conduct and other topics, run by the professional association. They are expected to follow similar activities during their career, but this is not yet compulsory.

The academic programs will probably be modified in a few years' time because of new European guidelines concerning actuarial studies. As a result, actuarial studies in Belgium may become two years at Master's level, and students with a Bachelor's degree (three years of university study) will be admitted to the program. It will then take five years at university to become an actuary instead of the current six to seven years.

The *Groupe Consultatif* representing the national professional associations of the Free European Exchange zone (so-called "Espace Economique Europeen," larger than EU) has set up a mutual recognition agreement. (*See* Appendix C.) A crucial step is the creation of a European program for actuarial studies. But this is for European actuaries (in the broad sense). For overseas actuaries, the rule is that they must apply to the Education Committee of the Belgian Society of Actuaries for recognition of the equivalence of their credentials to the Belgian requirements.

before attaining full professional status. The aim of this type of thesis is therefore to encourage the development of that experience and foster innovations in actuarial science. The Foundation for Promotion of the Actuarial Profession actually encourages this type of work by providing financial support.

Finally, candidates must have completed at least one year's practical experience in an insurance company or have done equivalent work. Experience may count as equivalent if it consists of practical applications actuarial methods, under the supervision of a Fellow of the Society of Actuaries.

Continuing professional development is not mandatory. The Society offers voluntary seminars and courses on topical issues when legislation is changed or the environment is changed otherwise. Most actuaries attend these activities. Another popular form of professional development is participation in actuarial conferences.

France

The education of actuaries in France is university-based. Three universities offer degree programs in actuarial science: Brest, Lyon, and Strasbourg. According to Morgan (*see* Reference 15, Appendix F), "The profession is still underdeveloped compared to the United Kingdom, and France is the only European country where actuaries are not a legally recognized profession." Morgan points out that "as in many European countries, the actuarial profession has been more academic and less practical than that in the United Kingdom, but this is changing as elements of accounting, law, and tax have been added to the course of study. These days, actuaries work in banks and consultancies as well as in insurance companies. In insurance their role is widening to include marketing and communication as well as just technical matters such as ALM [Asset and liability management], and embedded values are starting to become more widespread."

Germany

Germany has its own version of professional accreditation. In order to qualify for membership in the *Deutsche Aktuarvereinigung* (Actuarial Association of Germany), candidates must pass examinations testing their general and specific competence in actuarial science. The *Deutsche Aktuarvereinigung* has joined forces with the *Deutsche Gesellschaft für Versicherungsmathematik* (German Society for Insurance Mathematics) and the *Institut der Versicherungsmathematischen Sachverständigen* (Institute of Experts in Insurance Mathematics) and founded the *Deutsche Aktuar-Akademie* (DAA) (German Actuarial Academy), which provides basic and advanced training for actuaries. The DAA holds seminars and workshops for the courses in which actuarial candidates are examined.

The German accreditation system consists of three levels of examinations, each consisting of several courses. Each level is considered to require one year of preparation. Level 1 consists of three examinations and one compulsory course in data processing. The subjects examined include mathematics of the life insurance, mathematics of finance, and other elementary actuarial topics. Level 2 consists of two examinations, chosen from four topic areas: P/C, pensions and stochastic methods, real estate, and health. Level 3 consists of a compulsory seminar and examination in one of the following specialties: life insurance, P/C, pensions, applications of stochastic methods, health, and finance. Several German universities offer degree programs in actuarial science. Among them are the universities of Ulm and Göttingen.

Hong Kong

The Actuarial Society of Hong Kong does not conduct its own set of actuarial examinations at the moment. It relies on the exam systems of other established overseas actuarial bodies. Typically, to be admitted as a Fellow member of the Actuarial Society of Hong Kong, the member must be a Fellow of one of the actuarial bodies of Australia, Canada, United Kingdom, or the United States, although there is an increasing number from other countries, especially from Europe. Under the Hong Kong government's insurance company (qualification of actuaries) regulations, the qualifications for appointment as an Appointed Actuary are: Fellow of the Institute of Actuaries of England, Fellow of the Faculty of Actuaries in Scotland, Fellow of the Institute of Actuaries of Australia, or Fellow of the Society of Actuaries of the United States of America.

Until recently, students who wished to pursue an actuarial science degree had to travel overseas. Now three universities—the University of Hong Kong, the Chinese University of Hong Kong, and the Hong Kong Polytechnic University—offer a range of actuarial subjects.

India

The actuarial education in India is profession-based. The Actuarial Society of India offers a series of examinations that must be passed to qualify as an actuary. The Society was established in 1944 to provide a central organization for actuaries in order to raise the standards of competence and level of recognition of the actuarial profession. The structure of the Society and its examination syllabus are comparable to that of the Institute of Actuaries of the United Kingdom. As in other countries, actuaries in India are involved in insurance, pensions, investment, financial planning and management. According to the Society, actuaries have "an unlimited scope in countries outside India where the necessary infrastructure already exists to absorb them in suitable avenues like life and general insurance, operations research, statistics, investment, demography, etc. The remuneration offered is very lucrative and the job satisfaction is tremendous."

Ireland

The equivalent of a Fellow of the Society of Actuaries is a Fellow of the Society of Actuaries in Ireland (FSAI). Most Fellows qualify through the Institute or Faculty of Actuaries in the United Kingdom. Under the constitution of the Society of Actuaries in Ireland, all Fellows must be Fellows of the Institute or Faculty in the United Kingdom or via the various mutual recognition agreements such as the *Groupe Consultatif*, the Australian Institute, the Society of Actuaries, or the Canadian Institute of Actuaries. A foreign actuary can join the Society via a mutual recognition agreement with one of the aforementioned bodies.

Israel

The education of actuaries in Israel is concentrated in universities. In addition to courses in actuarial science available at the Hebrew University in Jerusalem and at the University of Tel-Aviv, the University of Haifa maintains an active research center in actuarial science, offering a Master's degree. The Israel Association of Actuaries is the professional body for actuaries in Israel and is a full member of the International Actuarial Association. One of its functions is to enhance the practical knowledge of graduates of the academic courses in Israel and abroad and examine these candidates for Fellowship. Individuals with actuarial training can also become qualified members of the Society of Actuaries of the United States by writing examinations at the permanent SOA examination center in Ramat Gan.

According to the historical account of the evolution of the actuarial profession in Israel [*See* http://hevra.haifa.ac.il], "the Israeli industry's approach to financial risk has consisted of adapting foreign solutions to better reflect Israeli reality and its needs. Thus, the mortality tables ... in use in Israel today come from England and are subject to an adjustment. However, this adjustment ... has no scientific justification or basis; at best, it represents the intuition of insurance company actuaries or, alternatively, it is a manifestation of the interests of such companies, a possibility that has drawn ample criticism. In the Western world, actuarial centers work to gather and analyze mortality data to provide the mortality tables necessary for performing precise calculations. In Israel, this step has yet to be taken. The Actuarial Research Center aims to close this gap."

Italy

Actuarial life in Italy is coordinated through the *Istituto Italiano degli Attuari* and by *Ordine Nazionale degli Attuari*.

The Institute is a member of the *Groupe Consultatif* and therefore has reciprocal agreements with the member countries of that group.

The Italian actuarial associations maintain a permanent professional development program through SIFA, *Corsi di Formazione Attuariale Permanente*, allowing its members to keep up-to-date with changes in actuarial practice resulting from globalization and European integration. The actuarial education in Italy is university-based and the title of fully qualified actuary is obtained through a state examination. To act as consultant an actuary must be enrolled in the National Register (*Albo Nazionale*), established by law in 1942.

The program of the state examination is under review to be consistent with the recent reform of the university system.

Japan

The actuarial education is profession-based. The Institute of Actuaries of Japan offers actuarial courses that enable applicants to acquire basic knowledge and to prepare for qualification examinations. Actuarial courses are divided into two categories, basic and advanced courses. The basic courses are intended for students of the Institute, while advanced courses are aimed at persons who have completed the basic subjects.

To become an Associate member of the Institute, candidates must pass examinations in the following five basic courses:

1. Probability and statistics.
2. Basic principles and applications of life insurance mathematics.
3. Basic principles and applications of non–life insurance mathematics.
4. Basic principles of pension mathematics and pension finance.
5. Basic principles of accounting, economics and investment theory. After passing these courses, candidates qualify for Associate membership in the Institute of Actuaries of Japan.

To become a Fellow of the Institute, Associates must pass two additional advanced courses: (LI1) Life insurance products and development and (LI2) Life insurance accounting, settlements of accounts, or (NLI1) non–life insurance products and development and (NLI2) non–life insurance accounting, settlements of accounts and asset management, or (P1) Tax qualified pension plan scheme and pension-related tax and accounting and (P2) Public pension system and employees' pension fund scheme. Fellowships are approved by the Board of Directors of the Institute. New fellows are also strongly recommended to take a half-day professionalism course.

The education system of the Institute is under review with the following objectives: broader areas to be examined and the completion of a professionalism course for fellowship will eventually be required.

Several Japanese universities offer courses on actuarial mathematics and risk management, but there are no exemptions for qualification examinations. In 2001, the membership of the Institute was made up as follows: 958 Fellows (including six honorary members), 772 Associates, and 1667 Students.

Malaysia

The actuarial profession in Malaysia is represented by the Actuarial Society of Malaysia. The Society does not have its accreditation process. Actuaries meeting the following criteria may be admitted as Fellows of the Society: (a) The candidates are Fellows of the Institute of Actuaries of England, the Faculty of Actuaries of Scotland, the Society of Actuaries of America, the Canadian Institute of Actuaries, or the Institute of Actuaries of Australia. Admission to the Society must be approved by the Executive Committee of Society. Qualified actuaries are allowed to practice in Malaysia if they reside in Malaysia or, in the opinion of the Executive Committee, are familiar with Malaysian conditions, and have paid the requisite admission and annual membership dues.

Fellows of the Society can become Appointed Actuaries of insurance companies by being approved by the regulatory authority in Malaysia (Bank Negra Malaysia). Appointed actuaries must be residents of Malaysia and have at least one year of relevant work experience with a Malaysian insurer.

Mexico

The education of actuaries in Mexico is university-based. To be able to work as an actuary in Mexico, and to be allowed to use the designation "actuary," candidates must fulfill three requirements: (1) They must complete a four-year undergraduate program in actuarial science which include 480 hours of unpaid socially valuable work. The Mexican syllabus is close that prescribed by the SOA. In fact, many students in Mexico are encouraged to write the SOA examinations. (2) They must write a relevant dissertation. (3) They must defend the dissertation before an examination committee.

Some universities accept graduate work in relevant academic programs and the passing of written and oral comprehensive examinations as dissertation equivalents. In order to be accredited as actuaries with signing privileges, graduates must have their university degrees approved by the Ministry of Education and obtain from the Ministry a *cédula profesional*. Certified actuaries are publicly sworn to uphold the code of ethics of the profession, but they are not required to become members of the Colegio or any other association of actuaries. A significant number of actuaries in Mexico work in non-traditional areas such as finance, government, planning, and information technology.

Mexico has two actuarial organizations: In 1962, the *Asociación Mexicana de Actuarios del Seguro de Vida* was formed. Its members tend to work in life insurance. In 1980, the association expanded its membership to include all actuaries and become the *Asociación Mexicana de Actuarios*. In addition, the profession established College of Actuaries in 1867, the *Colegio de Actuarios de México,* which was transformed into the *Colegio Nacional de Actuarios* in 1982. Membership in the College is not required to function as an actuary. Mexico has close to 2,000 actuaries, most of whom work in the Mexico City area.

Netherlands

The Dutch equivalent of the Fellowship of the Society of Actuaries is a Fellowship in the *Actuarieel Genootschap*. Two roads lead to this Fellowship:

Successful completion of the actuarial program of the *Actuarieel Instituut*. This involves between eight and nine years of study.

The other option is to complete a Master's program in actuarial science at the University of Amsterdam. This involves between four and five years of study, together with a successful completion of the two-year post-Master's course of study administered by the *Actuarieel Instituut*.

The Netherlands has a reciprocal agreement with the *Groupe Consultatif* for recognizing each other's Fellowships.

Norway

The education of actuaries in Norway is university-based. Since 1916, the University of Oslo has offered a degree program in actuarial science in insurance mathematics and statistics. In the 50s and 60s, a program of actuarial studies based on stochastic principles was established. Risk theory and non–life insurance were added to the curriculum in the 70s. Now, actuarial students are required to complete a five-year Master's program in mathematical statistics with specialization in insurance mathematics. Students must also write a Master's thesis equivalent to one-year of full-time work experience. The study of economics is no longer a required part of the program. The current thinking is to make the program more applied and introduce a business component. Since 1997, the University of Bergen also offers a degree program in actuarial science. Almost all actuaries in Norway belong to the Norwegian Actuarial Society.

Portugal

The Portuguese equivalent of a Fellow of the Society of Actuaries is an *Actuário Titular,* a full member of the Portuguese Institute of Actuaries. The education of actuaries is university-based. The Technical University of Lisbon offers courses in actuarial science. A graduate with a university degree in mathematics, economics, and management, together with appropriate course in actuarial science and three years of experience working as an actuary, can become a member. The candidate must prepare a report under the supervision of an accredited member of the Institute as part of the accreditation process. Foreign-trained actuaries must be have their professional credentials recognized by the Institute to become members.

South Africa

The life insurance industry was brought to South Africa by the British Settlers in 1820, and the first actuaries in South Africa were all Britons, employed by UK-based companies. As a result, South Africans who were interested in the actuarial profession were exposed to the UK Institute and Faculty of Actuaries. To this day, the vast majority of South Africans qualify via the UK actuarial education system. Applications for admission as a student member of one of the UK organizations are dealt with by the Admissions Committee of the Actuarial Society of South Africa.

Various South African universities offer undergraduate and postgraduate courses in Actuarial Science. Students taking these courses may obtain exemptions from most of the subjects required by the UK syllabus, depending on their university results. These exemptions make it possible for a student, after university, to have to write only one subject in the 300 series and the 400 series before qualifying as an actuary. A small number of South African students follow the courses prescribed by the US Society of Actuaries and the Casualty Actuarial Society. Examinations of the UK and US organizations are administered by the Actuarial Society of South Africa (ASSA) at centers in Cape Town, Durban, Johannesburg and Windhoek (Namibia).

Since September 2003, students have the option of writing a Fellowship paper based on South African legislation and regulation, instead of the UK paper. The local-content paper is set in conjunction with the UK actuarial education authorities and results in the same qualification being awarded, i.e., either FFA or FIA.

Attendance at the ASSA professionalism course is required for all actuaries within a year of completing the exams of either the Institute or Faculty of Actuaries. The course is recognized by both the Institute and Faculty of Actuaries. The course structure is based closely on that used for the professionalism course run by the Institute and Faculty in the UK. To a large extent, the course material is identical to that used in the UK.

The aim of the ASSA Professionalism Course Committee was to construct a course that meets the Institute and Faculty of Actuaries requirements for recognition while providing sufficient local South African content to keep the course relevant and interesting for South African delegates.

In addition to the incorporation of case studies reflecting issues facing the profession in South Africa, the UK course material is supplemented by course material on ethics, as used by the Institute of Actuaries of Australia. The course material is further supplemented by locally developed material on legal liability and conflicts of interest.

ASSA's professionalism course is run twice a year. It is run over two days on a residential basis and comprises a total of around 12 hours of working time. The course is run by two lecturers (both experienced qualified actuaries) and a guest speaker (also an actuary). The guest speaker is chosen to talk about one of the wider fields (healthcare, general insurance or investment work—depending on the number of delegates involved in each of these wider fields) and to illustrate issues of a professionalism nature facing actuaries in such a field. The course utilizes a variety of media, including lectures, group workshops, case studies, newspaper clippings, video and audio material.

ASSA provides regular opportunities for its members to keep up with technical and professional developments. A Continuous Professional Development Compliance Certificate is mandatory for actuaries who are active in certain fields.

The Financial Services Board, as regulatory authority of the financial services sector in South Africa, approves actuaries as statutory actuaries of life offices and as valuators of retirement funds. In this process, the Board requires an applicant to submit a relevant practicing certificate, which is issued by ASSA.

South Africa has 497 actuaries and just more than 1,000 students. Students have been qualifying at a rate of between 50 and 60 per year for the past three years.

South African law stipulates that valuations of life offices and retirement funds have to be performed by actuaries, but there is no formal, statutory actuarial involvement in other fields at present. Some progress has been made with regard to the formal involvement of actuaries in healthcare and short-term insurance.

The major fields of activity for actuaries practicing in South Africa are life offices (31%), employee benefits (28%), consultancy work (involving life insurance, healthcare and employee benefit work, but not employed by an insurer—21%), healthcare (9%), and investments (5%). The remaining 6% are involved in short-term insurance, academia, etc.

Spain

Since the establishment of the *Instituto de Actuarios Españoles*, the Spanish professional association, and still nowadays, the only requirement to be admitted as a full member is to have the actuarial and financial sciences degree. This higher education degree, offered by several universities through their faculties of economics and business administration, is a full actuarial education program, which normally takes two years and consists of 150 credits (one credit involves ten effective lecture hours), where almost half of the credits must be in the following subjects:

Actuarial statistics (including topics on stochastic processes, survival models and, partially, risk theory), financial mathematics (also including topics on investment), actuarial mathematics (including risk theory and life and non–life insurance mathematics), accounting and financial reporting in insurance, banking and investment, insurance, banking and stockmarket regulations, social security economics and techniques.

In deciding on the rest of the program or syllabus, each university has a significant degree of autonomy, but most of them expand the number of credits in actuarial mathematics, financial mathematics, statistics, accounting and financial reporting, and then offer specialized courses in private pension plans, financial instruments and markets, taxation, solvency, reinsurance, insurance and financial marketing, computing, and so on.

Students who want to pursue such a program must have an undergraduate degree, usually in economics or business administration. It must include courses in mathematics, probability and statistics, economics, finance and accounting, and financial reporting.

However, significant changes in the Spanish university education system will be made to meet the Bolognia agreements within the European Union. As a result, the actuarial and financial sciences degree program is likely to undergo relevant changes, either as an undergraduate program or a graduate program or both.

In addition, the *Instituto de Actuarios Españoles* is working toward implementing the changes required to meet the education requirements set by the European Union Recognition Agreement (*Groupe Consultatif* Core Syllabus). This program and the examinations involved, will be set and run by the Spanish Institute of Actuaries through the "Actuarial Training School."

Sweden

In Sweden, the education of actuaries is based on a hybrid system. Membership in the Svenska Aktuarieföreningen is granted to individuals who have fulfilled appropriate academic requirements. (1) Candidates must have the necessary grades in basic mathematics and mathematical statistics. (2) Candidates must also have obtained an actuarial diploma, granted to a person who, in addition to what is required for step one, has the necessary grades in actuarial subjects, including actuarial science, law and economics, have written an approved actuarial paper, and have at least two years of experience working as actuaries. The required knowledge and experience may have been acquired in another country, and the Svenska Aktuarieföreningen is a member of the European cross-border mutual recognition agreement. License to act as an Appointed Actuary is granted by the

Finansinspektionen supervisory board. The requirements are similar to those required for the actuarial diploma.

Switzerland

In Switzerland, the profession is made up, as everywhere else, of actuaries with university degrees who have passed special professional examinations and have special legislative powers assigned to them in their capacity as general insurance and life insurance actuaries. In the case of pension actuaries, the situation is even more structured. Pensions in Switzerland are subject to special laws and pension contributions are mandatory for all employers. This is an important difference between Switzerland and North America. As a result, pension actuaries must obtain additional qualifications to practice.

Swiss Actuaries receive their qualification from the Swiss Association of Actuaries and hold the title "Actuary SAA." They are qualified in the sense of being full members of the IAA. If Fellows of the SOA want to work for pensions funds in Switzerland, they must pass an appropriate special examination, in addition to being qualified as an Actuary SAA.

The academic and professional steps required for Fellows of the Society of Actuaries to be recognized as an Actuary SAA, involve an evaluation of their credentials by an admissions commission of the SAA. They will decide whether there are deficiencies to be filled by extra examinations or whether all aspects of the Swiss syllabus have been covered. In the first case, the candidate will be notified of the exams to pass, in the second case the admission is straightforward.

If Fellows of the Society of Actuaries want to be able to give statements to Pensions Funds, they must submit their credentials to the examination commission of pensions funds experts. Afterwards the procedure is the same as for the Actuary SAA, but there will, with great probability, be a deficiency in legal knowledge. Fellows of the SOA will therefore almost certainly have to pass at least the legal examination before receiving their state diploma.

Actuarial studies in actuarial mathematics in Switzerland involve completing a course of studies based on the Swiss syllabus. Certain Swiss universities are accredited by the SAA. An appropriate degree from these universities means that the academic requirements have been met. For students from other universities, the same procedure as for a Fellow of the Society of Actuaries is required: existing knowledge is evaluated and deficiencies have to be made up for by special examinations. In addition, actuaries must have three years of practical experience and have passed the examination colloquium.

For pension fund experts there are special admission examinations, preliminary examinations, and a comprehensive examination, all of which must be passed, with the possibility of taking into account former qualifications.

United Kingdom

The Institute of Actuaries in England and the Faculty of Actuaries in Scotland are the two professional bodies for UK actuaries, working closely together as *The Actuarial Profession* in the United Kingdom. Upon qualification members become either a Fellow of the Faculty of Actuaries (FFA) or a Fellow of the Institute of Actuaries (FIA).

Members have to join the profession as student members in order to sit the professional examinations. The minimum entrance requirement is a B in A-level in mathematics or equivalent. However 95% of all entrants are university graduates. Although any degree is acceptable, most actuaries possess either a first- or second-class degree in a mathematics-based field. For holders of a second-class honors degree or above, the mathematics A-level requirement is a C. For those with a mathematics or actuarial science honors degree, mathematics A-level is not required.

In order to become a Fellow, actuaries have to pass the professional examinations and for the Institute (but not the Faculty) gain three years' relevant work experience. On average, qualification takes at least three years. Student members take the examinations at their own pace, whilst working for an actuarial employer (probably an insurance, consultancy or financial organization).

The actuarial education system in the United Kingdom has the following components:

1. Diploma in actuarial techniques. This is awarded to candidates who have passed all the subjects in the 100-series of the professional examinations or have gained exemptions through designated university courses.
2. Certificate in finance and investment. This is awarded to candidates who have passed subjects 102, 103, 107, 108, 109 and 301 investment and asset management. It demonstrates knowledge in the investment area.
3. Associateship of the Institute or Faculty (AIA or AFA). Associateship is awarded to candidates who have gained the Diploma in actuarial techniques, passed subject 201 Communications and the four 300-series subjects, and attended an Associate professionalism course. The subjects studied cover both the theoretical foundation for actuarial practice and the principles behind actuarial applications work.
4. Fellowship of the Institute or Faculty (FIA or FFA). The Fellowship is the main qualification as an actuary. It is awarded to candidates who have passed the examination requirements for the Associateship and in addition have passed one actuarial subject at Fellowship level chosen from investment, life insurance, general insurance or pensions. Candidates have to demonstrate that they can apply the theoretical framework in an established practical country-specific applications area of actuarial work. A Fellowship professionalism course must be attended within one year of qualification.

A significant number of local societies form the lifeline of the actuarial profession in the United Kingdom: Staple Inn Actuarial Society, Birmingham Actuarial Society, Bournemouth Actuarial Society, Bristol Actuarial Society, Channel Islands Actuarial Society, Faculty of Actuaries Students' Society, Glasgow Actuarial Students' Society, Invicta Actuarial Society, London Market Actuaries Group, London Market Students' Group, Manchester Actuarial Society, Manx Actuarial Society, Norwich Actuarial Society, Society of Actuaries in Ireland, White Horse Actuarial Society, Yorkshire Actuarial Society.

The United Kingdom has reciprocity agreements with the countries belonging to the *Groupe Consultatif Actuariel Europeen*. Some other countries, such as Hong Kong, use the UK process to certify actuaries.

Missing Countries

The list of countries covered in this section is obviously incomplete. Most countries around the world have insurance companies and either privately run or public pension schemes. Actuarial considerations are therefore relevant to all countries. Rather than being encyclopedic, the choice of countries profiled in this section is intended to give you an idea of the variety of different national traditions and models for being an actuary. It also shed some light on international mobility.

In his article on the state of actuarial science in China (*see* Reference 2, Appendix F), Alexander explains that China and other Pacific Rim countries are, at this point, *actuarially emerging countries*. We therefore refer to other sources for information about the state of the actuarial profession in these countries. A similar remark applies to Russia and many countries in Eastern Europe, the Middle East, Africa, and Central and South America. License to practice in emerging countries is often based on a mix of university education, working experience, and professional recognition by designated government agencies. Here are three typical examples from Eastern Europe:

In Croatia, an accredited actuary must be full member of the Croatian Actuarial Association. To become accredited, candidates must have at least two years actuarial work experience, have successfully completed a program of actuarial study which follows the guidelines of the International Association of Actuaries and is recognized by the Assembly of the Croatian Actuarial Association.

An example of a recognized program of study is the Master's program in actuarial mathematics offered by the Department of Mathematics of the University of Zagreb, together with examinations set by the Croatian Actuarial Association.

Accredited actuaries must be Croatian citizens, have a degree in economics, mathematics, physics or engineering, and have passed the examinations set by the Ministry of Finance or equivalent. An example of a recognized program of study is the actuarial program jointly organized by the Croatian Actuarial Association, the Department of Mathematics of Zagreb University, and the Actuarial Department of the Government of the United Kingdom.

To become accredited pension actuaries, candidates must first become accredited actuaries as described, have at least three to five years' experience in actuarial work (depending on their specialization at the undergraduate level), have a postgraduate degree in actuarial mathematics which includes pension insurance, or have an equivalent actuarial education from abroad, recognized by the Croatian Actuarial Association. In addition, they must have passed a special examination set by the Agency for Supervision of Pension Funds and Insurance.

In the Czech Republic, a permanent commission of the Czech Society of Actuaries, approved by the government, issues licenses to practice. Candidates must meet academic requirements based on the Czech tradition in actuarial education. The certification process also relies on the criteria for actuarial practice established by the *Groupe Consultatif*. In accordance with the Insurance Act, actuaries in the Czech Republic can become Appointed Actuaries upon approval by the Ministry of Finance. Foreign-trained actuaries must be fully qualified members of an actuarial society that is a full member of International Association of Actuaries.

In the Slovak Republic, anyone with an actuarially oriented university degree and one year of relevant work experience can become a member of the Slovak Society of Actuaries. To become an Appointed Actuary, Slovak Insurance Law requires that candidates have an appropriate university-level education, have passed a special examination and have three years of relevant practical experience. The Slovak Financial Market Authority organizes the examination. Candidates are not required to be members of the Slovak Society of Actuaries.

Chapter 2

ACTUARIAL EDUCATION

2.1 The IAA Syllabus

The International Actuarial Association surveyed its members about their educational practices. It identified a number of academic topics as describing the repertoire of scientific knowledge and competency areas of actuaries. The list of topics is, in a sense, more important than the answers to the survey. It summarizes the understanding of leading actuaries of the core tools of their profession. Here is the list, which divides actuarial education into 10 broad areas:

Financial Mathematics

Aim To provide a grounding in the techniques of financial mathematics and their applications.

Topics Introduction to asset types and securities markets; interest, yield and other financial calculations; investment risk, introduction to stochastic interest and discount; market models, e.g., term structure of interest rates and cash flow models.

Probability and Mathematical Statistics

Aim To provide a grounding in probability and mathematical statistics.

Topics Concepts of probability; random variables and their characteristics; methods and properties of estimation; correlation and regression analysis; hypothesis testing and confidence intervals; data analysis.

Economics

Aim To provide a grounding in the fundamental concepts of both microeconomics and macroeconomics.

Topics Microeconomics; macroeconomics.

Accounting

Aim To provide the ability to interpret the accounts and financial statements of companies.

Topics Basic principles of accounting—including the role of accounting standards; fifferent types of business entity; basic structure of company accounts; interpretation and limitation of company accounts.

Modeling

Aim To provide an understanding of the principles of modeling and its applications.

Topics Model structures; selection process; calibration; validation; scenario setting; sensitivity testing; limitations.

Statistical Methods

Aim To provide the skills and expertise in the use of models appropriate for the understanding of risk in a range of actuarial work.

Topics Statistical models, such as regression and time series; survival and multistate models; risk models (individual and collective); parametric and nonparametric analysis of data; graduation principles and techniques; estimation of frequency, severity and survival distributions; credibility theory; ruin theory.

Actuarial Mathematics

Aim To provide the skills and expertise in the mathematics that are of particular relevance to actuaries working in life insurance, pensions, health care and general insurance.

Topics Actuarial mathematics as applied to life insurance, pensions, health care and general insurance; types of products and plans—individual, group and social insurance arrangements; pricing or financing methods of products and plans; reserving; reinsurance.

Investment and Asset Management

Aim To develop the ability to apply actuarial principles to the valuation, appraisal, selection and management of investments.

Topics The objectives of institutional and individual investors; types of investment (bonds, shares, property and derivatives); regulation and taxation of investments; valuation of investments; portfolio selection—incorporating assessment of relative value; performance measurement; portfolio management.

Principles of Actuarial Management

Aim To develop the ability to apply the principles of actuarial planning and control needed for the operation of risk related programs on sound financial lines.

Topics The general operating environment; assessment of risks; product design and development; pricing and assumptions; reserving and valuation of liabilities; asset and liability relationships; monitoring the experience; solvency of the provider; calculation and distribution of profit (surplus).

Professionalism

Aim To develop awareness of professionalism issues and the importance of professionalism in the work of an actuary.

Topics Characteristics and standards of a profession; code of conduct and practice standards; the regulatory roles of actuaries; the professional role of the actuary.

This list can be found in the *Education Syllabus* section of the website of the International Actuarial Society. It is a wonderful conceptual organizer for the overwhelming mass of mathematical, economic, financial, and other ideas that make up the syllabus upon which the SAO and CAS examinations are based. As you read on, you might try to fit the listed topics and sample examination questions into this scheme. It will help you with the conceptual order and organization of the material that follows.

2.2   The SOA and CAS Examinations

In Section 2.1, we saw that there are major differences in the actuarial education around the world. However, the SOA and CAS qualifications are respected and honored around the world. Twice a year, in May and November, students can write SOA and CAS examinations in multiple centers in the United States and most provinces in Canada, as well as in international examination centers. If you peruse the SOA, CAS, and CIA website, you will see a long list of biannual

American and Canadian examination centers. But the list of international centers is equally impressive. Here, for example, is the list of permanent international test centers outside the United States and Canada, where you are able to write SOA and CAS examinations:

Accra (Ghana), Athens (Greece), Bangkok (Thailand), Beijing (China), Bogota (Colombia), Bridgetown (Barbados), Buenos Aires (Argentina), Cairo (Egypt), Capetown (South Africa), Changsha (China), Colombo (Sri Lanka), Delhi (India), Guangzhou (China), Hamilton (Bermuda), Harare (Zimbabwe), Hefei (China), Ho Chi Minh City (Vietnam), Hong Kong, Hyderabad (India), Jakarta (Indonesia), Johannesburg (South Africa), Karachi (Pakistan), Kingston (Jamaica), Kuala Lumpur (Malaysia), Lagos (Nigeria), Lahore (Pakistan), Madrid (Spain), Manila (Philippines), Mumbai (India), Nairobi (Kenya), Nassau (Bahamas), Oxford (England), Panama (A Mundial), Paris (France), Port-of-Spain (Trinidad), Ramat Gan (Israel), Santiago (Chile), São Paulo (Brazil), Seoul (South Korea), Shanghai (China), Shenzhen (China), Singapore, Sydney (Australia), Taichung (Taiwan), Taipei (Taiwan), Tianjin (China), Tokyo (Japan), Warsaw (Poland), Xi'an (China), and Zurich (Switzerland).

This list certainly shows you the high regard in which the North American system of actuarial education is held around the world. In places where no permanent examination center exists, arrangements can often by made by candidates by finding their own supervisors of examinations acceptable to the Society of Actuaries. Supervisor must be either members of the Society of Actuaries, members of the Casualty Actuarial Society, members of the Institute of Actuaries (England), members of the Faculty of Actuaries (Scotland), or be a tenured academic or other qualified testing professional. If no such supervisor is available, approval may even be given for writing the examinations at an Embassy of the United States.

General Comments

If you take a closer look at the May 2001 examinations in Courses 1 through 4, you will discover that most of them deal with some aspect of mathematics and statistics in a business context. In order to pass these examinations, you must therefore have a solid understanding of all three subjects. The answers to most questions in Courses 1, 3, and 4 involve relatively short calculations. Course 2 is slightly different. In addition to being able to carry out mathematical and statistical calculations, you must be able to understand the definitions and interrelationships of concepts from economics and finance. You will also notice that many questions involve both definite, indefinite, and multiple integrals, as well as ordinary and partial derivatives. Hence, a good command of calculus is essential. Exponential and logarithmic functions are core functions, and so are geometric progressions. You will also discover that indispensable concepts from statistics are probabilities, distributions, random variables, and expected values. So are mean and variance. You will also notice that among the probability distributions, the Poisson distribution comes up most often. While the specific questions will obviously vary from year to year, it is clear from the nature of actuarial science that the mentioned ideas and techniques from mathematics, business, and statistics will always be part of the skills an actuary is required to possess.

Theory and Practice

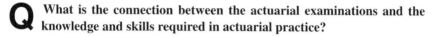 **What is the connection between the actuarial examinations and the knowledge and skills required in actuarial practice?**

Answer Not as important as I expected, especially the first four exams, which are very different from actuarial practice and so are the required skills. The examinations in Courses 5 and 6 are closer to the real world, and I have heard that the examinations in Courses 7 and 8 are much more like real consulting situations, although you can't really have a consulting situation in an exam. My overall feeling is that the more you advance in your exams, the more related they are to real life. I just don't feel that the mathematics background for the examinations in Courses 1 through 4 is that important in real life. I am not saying it is not important, just that it is not a major part of success in the workplace in the first few years of employment.

Answer Limited. Material is either too theoretical, or off whatever is required from us in the day-to-day life.

Answer I would say that the actuarial examinations go much more deeper. We have to know every little formula, even if it applies only in a unusual situation. In our job, we can use one formula and adapt it to given situations. Also, the Associateship exams touch every field (pension, insurance, finance, etc.). So an actuary needs to be versatile.

Answer Actuaries need to follow specific methods that are regulated in order to make sure everyone follows the same standards. The actuarial examinations are a way to introduce the various methods of calculating reserves, and they give you a background in subjects used in real life situation.

Answer Actuarial exams helped me to develop my capacity to focus on problems and to solve them. Since I can solve a lot of problems, I do not need to constantly refer to books. Also, actuarial examinations help us learn how to work really hard and well.

Answer The knowledge acquired in the study for actuarial examinations serves as a good learning base for the actuarial practice. As in all professions, most of the learning is done on the job, but in the case of actuaries, since such an extensive knowledge must be acquired, the examinations form the practicing actuary.

Answer Courses 5 to 8 are more directly applicable to work, although Courses 2 and 3 are often directly used too. A strong basis in the fundamentals of Courses 1 through 4 is needed to do Courses 5 through 8. Some topics only apply to certain jobs and the degree of applicability is job-specific too. I always treated Courses 1 through 4 as first- or second-year university courses. This puts students from various educational backgrounds on the same common basis from which they can build up their knowledge. Courses 5 through 8 are more like 3rd- or 4th-year university courses—higher level of knowledge, more practical, some specialization occurring.

Answer The mathematical exams (Courses 1 through 4) are directly related to the actuarial needs. Other exam material is necessary, but the way it is tested (i.e., learning everything by heart) is not related to actuarial needs...

Answer Not much when it comes to learning things by heart part, because you have all the documentation handy, especially with the net. The first part (mathematics), though, is really good.

Answer I see the applications of life contingencies daily in my pension work.

Answer The actuarial exams expose the student to some of the available literature in the field. The actuary may refer back to some of these sources later.

Answer Actuarial examinations provide a very technical introduction to the skills required in actuarial practice. Most of the formulas memorized in the early mathematical exams will hardly be used in practice. Knowing how to apply the concepts presented in the examinations along with the skills outlined in the previous question are really what are required in actuarial practice.

Answer Actuarial examinations will teach the actuary all the required actuarial skills. They are necessary. They will also teach (at least for CAS exams) about the insurance business. But they will not teach business skills. Basically, on the P/C side, exams teach how to price products, how to calculate IBNR (incurred but not reported) loss reserves, how to read a financial statement, how to value investments, how to perform modeling, and also teach the basics of P/C insurance (products, concepts, etc.). I would say all exams are relevant to actuaries.

Answer CAS only: there is a very limited connection between the exams and the practice, except for the more advanced CAS exams (mainly Parts 5, 6 and 7; less so with Parts 8 and 9).

Writing the Examinations

Q In what order did you write the SOA or CAS examinations? Explain why.

Answer I wrote the examinations in the order 1, 2, 3, 4, 5. I thought if it was done this way, I should write them in that order!

Answer In the normal order. Did the mathematics part while in University, and the rest while working as it combined a little better like that.

Answer I did the following: 1, 2, 3, 4. Followed by 6, as it is offered once a year (spring), and then 5 in the fall of the same year. Then I did 7 and 8 the following year. TIP for those writing Exam 8 (Pensions): if I could do things over again, I would have written 8P at the same time as 5. There is a lot of overlap between 5 and 8P, and writing them together gives you a "free" attempt at 8P (if you fail, you can write it the next year with 8R, if you pass, your next fall sitting will be easier).

Answer So far, I've written them in order. I've done the first four because they are all the same type (multiple-choice only). I thought I needed more time to prepare to the others. But if I would fail one many times, I would skip it and try another one, to clear my head a little bit and help get motivated again.

Answer So far, I have written: May Course 1, May Course 3, November Course 2, and I am planning to write Course 4 next May.

Answer The suggested order (1, 2, 3, 4, etc.) because I had no reason to deviate.

Answer I wrote the first three exams in the order 1, 2, 3. When I wrote the second one, I didn't have all of the university courses required in finance to pass the exam, but I decided to study this part by myself. It worked well, but it is obvious that it is harder when you have not taken all of the required courses. However, since I had do study for the more advanced exams by yourself in any case, it was good practice to learn to study by yourself. There is no exact reason why I wrote the exams in the given order. I just felt that way.

Answer I have only written Course 1, but plan on going about them 1, 3, 4, 2, and then the rest. I want to write Course 2 last since I believe that the content of this exam differs too much from the 1, 3, 4, since in my opinion, these courses are more actuarial in nature.

Answer I wrote them in pretty much numerical order. I mostly took exams right after I had taken the appropriate university courses. I think my path would have been 1, 2, 4, 3, 6, 5, 7, 8 (hopefully I'll pass Exam 8).

Answer Normal order: Examinations 1 to 8.

Answer I wrote the exams in order. I felt it made more sense to do it this way. There are some candidates who choose not to do this. A few of the reasons are as follows. If a higher numbered exam is more relevant to some candidates' current work responsibilities, they may elect to write it before attempting a lower numbered exam. CAS Exams 5 and 9 both deal with ratemaking topics. Candidates who have successfully passed Course 5 sometimes wish to continue with the same topic for the next sitting. One of the main disadvantages of not writing the exams in order is the possible delay in obtaining your Associateship. In Canada, there is no real advantage to having your ACAS from a signing standpoint. However, it is still an achievement that is recognized and is often rewarded by employers.

Answer The order 1, 3, 2 best fit my university program.

Answer I don't remember, but basically in order. I took the life contingencies exam before the theory of interest because I thought it was more useful for the job I had at the time.

Answer I wrote the first four in order. Because of the timing of exams (some are only offered in the spring and others only in the fall), I sometimes wrote a more advanced part since it was offered at the next sitting. I also wrote Course 8 before Course 7. There's no reason to write the exams in order, it really doesn't make much of a difference.

Answer I wrote the exams under the pre-2000 system from 1 to 10, in order. If I failed one, say in November, I would write it again the following year, and keep progressing upward with the May sitting. There is no reason why, although it does make it easier since some of the later exams can rely on your knowledge of earlier exams.

Answer I wrote the SOA examinations. I wrote some of them under the pre-2000 system and some under the post-2000 system. I wrote the mathematics exams as soon as I had taken the related class, sometimes before taking the related class. I wrote the advanced exams in the order they were given. Advanced exams are only given once a year. Therefore, I wrote the exam that was offered at each sitting.

Answer For the first few exams, the order was mainly based on the order in which I took courses in school. For the "middle" exams, either no course was given at school or I decided to write the exams before taking the course. In such a case, I would look at the "syllabus of examinations" and try to see what I was comfortable studying for. For the more advanced exams, I wrote them in the order they were given (5, 6, 7, and so on).

Answer Always followed the standard order.

Difficulty of the Examinations

Q Which SOA or CAS examinations did you find to be the most difficult and why? Illustrate your answer with examples.

Answer I can't say I have found one more difficult than others, it is really the time to put into the study of the material that is more difficult. Of course, I found that Course 5 had much more new material than Courses 1 through 4, and it took more time to study for it, which I think is normal. Exam 4 was maybe more difficult to study for in the sense that it is a mixture of many small parts of material not really related one-another.

Answer Exams 4 and 6 were most difficult for me. Exam 4 requires a lot of memorization of formulas. It is the exam with the least useful content for someone working in my area (pensions). I also found Exam 6 difficult, probably mostly because it was my first written exam. The transition from multiple-choice exams to written exams is a difficult one. Written exams require radically different studying and test writing techniques.

Answer I've tried Exams 1, 2, 3 and 4. So far, the fourth one was the most difficult. The amount of formulas to learn by heart is overwhelming. You have to learn a lot of details to pass. And all the material was on subjects that I don't use at work.

Answer I struggled with SOA Course 6, because I had very little background in finance. Therefore, I needed to put in a lot of hours to make sure I understood the concepts. Course 8 was the hardest to date, because it involved less material for which you can study, and more experience-based concepts. For example, the exam contained questions with actual situations a consultant would have to deal with in day-to-day situations (i.e., the client sends an age service table and needs the actuary to calculate the cost of a benefit upgrade). These types of questions were not explicitly dealt with in the syllabus.

Answer Between the first three, I think that the hardest was the third one. Mostly because there is much more material in this one than in the previous one.

Even if I studied about 250 hours, I didn't have time to learn perfectly about topics covered in that exam. However, I still don't know my result, I cannot tell if it was really the hardest one until now. Also, I think the first one is hard but for different reasons. Since it is the first one, you have to learn to do calculation and to answer really quickly. Many personal skills need to be developed, for example, a good way to manage your stress.

Answer Course 6, because it is based on advanced finance principles that I don't use at work. My only background was finance courses at university.

Answer Cannot answer this question from personal experience (only one written!) but I tend to believe that everyone has their own weakness and therefore everyone will find different exams more difficult than another.

Answer They're all hard. The written answer exams were tough because of the amount of material that needed to be learned. I needed four months of solid studying to pass those. The mathematics exams didn't require as much, but it was different studying—old questions mostly. I only needed two to three months for those. Course 3 is difficult because it's got some really tough concepts to learn—and memorization doesn't work. You need to understand what is going on.

Answer Course 3 requires a lot of practice. Course 5 requires you to spend a lot of time learning lists by heart.

Answer Course 2 was the most difficult. The questions were not computational, but more theoretical. And even if I did know the answer to a question, it was not obvious which was the correct answer since many of them seemed correct.

Answer Course 8 (Life). The longest, the most material, the toughest conceptually. The exam included a 17-point question—very hard.

Answer I failed the theory of interest exam because I didn't memorize enough formulas, and ran out of time trying to develop everything from first principles. Once I memorized everything there was no problem.

Answer Part 7C, since it was the first one with Canadian content. I found it difficult because it was my first complete exam (the first five were partitioned when I wrote them) and also because I was still in university and had no experience in reserving and accounting.

Answer Course 5 for me was the most difficult because it is a very general exam that spans all practices: group insurance, casualty insurance, pensions, life insurance, etc. Since most people have worked only in one field, it is hard to grasp the concepts for all the other fields at once. There's almost too much information to know all at once. You find that you know the section in which you work quite well, but know nothing about the other sections. From a passing perspective, this also means that you will answer your questions quite well on the exam, while others will answer their section's questions quite well, leading all questions to have been very well answered by the people in those sections. This raises that passing bar compared to exams where everyone is in the same boat for all of the questions, and some questions may never be answered well by anybody.

Answer The most difficult, for me, was CAS Course 7, which deals with annual statement, taxation and regulation. The difficulty stems from the tremendous amount of minutiae one has to memorize. The exam, however, is important, especially for corporate actuaries who may have to deal with all of the above issues. Taxation in the projection of Financial Statements. Annual Statement understanding is also required for monitoring of company and competitors' results.

Answer Course 8. It is the longest exam (now six and a half hours long) and is track-specific. I found it particularly hard because at the time I wrote it, I had less than one year of experience. Therefore, the material included in the exam was all new to me. Moreover, being a Canadian, I am not familiar with the material used in the United States. There is a lot of US content in Course 8, and I believe it represents an additional difficulty for Canadian students.

Answer The CAS old Part 8 (whose material is now partly covered under Part 7C) was definitely the toughest exam I ever wrote. It took me four attempts (this is 4 years!!!) to succeed. This exam was extremely theoretical and very boring. It dealt with "law and insurance," "regulations of insurance," and "government insurance plans." There was more to it, but I do not remember everything. Most of the material was specific to the United States, rendering its learning very painful and rather useless. We had to read and memorize "court cases" (the actual court cases transcripts) (we all wondered what benefit there was behind learning these). Overall, most people I knew felt that learning this material was not the best use of our time. Fortunately, when the new exam system was implemented and Part 8C was torn to pieces, the CAS only kept the most relevant pieces and included them with the new Part 7. To sum things up, I had a hard time passing the exam because of the following: (1) The amount of material to learn (about 2000 pages from which questions were picked); and (2) My lack of interest in the topics covered because they were not relevant to Canada and also not relevant in general.

Answer CAS Part 7C because it involved memorization and no mathematical or numerical problems.

2.3 Ways to Pass Examinations

Q What were your study tricks and study processes that helped you pass your actuarial examinations? Illustrate your answer with examples related to the SOA and CAS examinations.

Answer Read the books twice, make my own summaries based on my readings and also the summaries available on the market. Read a lot and asks questions to fellow workers who have been there before.

Answer For the multiple-choice exams, do as many practice tests as you can get your hands on. I know of a lot of people who have made the mistake of not get around to doing the practice tests, because they haven't finished the readings. The practice tests are the most important things to read, try, re-read and review. If you can pass the practice tests, you will almost certainly pass the real test. The written answer tests require different preparation techniques. Many people find the first written answer test they write is a real adjustment. The best piece of advice I can give is to memorize lists of important items: even if a direct regurgitation question is not asked, knowing these lists will help you think of the important topics to cover while under pressure.

Answer Do a lot of exercises. I'm not good with just memorizing, I need to practice. So I've done as many exercises as I could. I also tried to vary the sources (not just from the ACTEX [study aids], for example). Start in advance so you don't feel rush at they end and panic. Set a goal (for example, study 300 hours overall), do a schedule and note your progress. One week before the exam, go through the books and write a summary sheet with the things you still struggle with.

Answer You need to find a partner that you will not necessarily study together, or be writing the same exam, but someone who will motivate you, and you will push each other to study, and to not procrastinate. At two months before the exam, it is easy to slow down your studying but this is the time that if you have someone else who is also studying, will not let you relax and take a break!!!

Answer For the first four SOA exams, I made sure that I understood every practice question that I came across. Practice, practice, practice. Course 5 was a crash course in memorization. I just wrote and re-wrote the study notes as much as possible until it sank in! Course 6 was a combination of learning financial and mathematical concepts and some memorization (although the understanding component is much more important). Course 7 was straightforward: show up at the seminar, follow instructions, and write an essay.

Answer I take a lot of time to study theory and to do practice exercises of the ACTEX [study aids]. When I finish it, I do sample examinations from previous years in real time. That way, I can have a real idea of my studying status. I think that the more secure way to pass an exam is to study many hours. For me, there are no other tricks.

Answer I used a lot of memory tricks. For example, in order to remember a whole list of items, I created a word with the first letter of each item of the list, etc. It also really helps to try to apply the concepts we learned (it is easier to remember when you understand the applications).

Answer Start early. In the case of a student going to school and writing exams, starting early is crucial if you do not want to be behind in either. For Course 1, other than starting early, going through the theory and practicing problems is very important. Doing the same problem over and over again is useful. And a couple of weeks before the exam, doing the sample exams found on the SOA website are extremely useful to familiarize oneself to the format and type of questions which you will be asked. I have found the same type of questions always come up and so understanding the sample exams is very important.

Answer Here is what worked for me:

▶ *Practice Exams.* On early exams, do practice exams under timed conditions, learn rest of material from answer guide, recognize what formulas had to be memorized.

▶ *Ask Others.* On later written exams, early failure led me to interview successful students on their study techniques, and select some that might work for me. My goal was to pass, not to learn the material better than anyone else.

▶ *Prepare Early.* Personally, I started early (January/July); a week after the results came out. My goal was one hour per day plus study time—I had a calendar at work where I couldn't ignore it, and put stickers on every day that I met my goal. It is especially important to motivate yourself with short-term goals early in the study schedule. When someone else at work was writing the same exam, I used their progress to pace myself through the material, although I never studied with anyone else. I took a local two-day course for CAS Part 5 (ratemaking), which scared me into studying even harder for the last month. Otherwise, I don't think much of courses—you can't depend on them to prepare you. I kept asking myself whether what I was doing was going to help me on the day of the exam, and stopped doing it if it wasn't (e.g., writing pretty notes, spending too much time on a problem I couldn't solve), and reminded myself that nothing else (e.g., how much time I had spent) counted when they marked the exam.

► *Make Notes.* I read the material on each paper while writing notes for anything I didn't think I would remember on the day of the exam (this helped me make sure I was absorbing the material, and not just skimming). I concentrated on writing lists and formulas, since it can't be on the exam if they can't make it into a question. I did questions from old exams following each paper to make sure I had learned the right material. In the final weeks before the exam, I reviewed my notes, forming the material into questions on the back of the prior page, so I could study by covering the answers and asking myself the questions. I also did timed exams from prior years. I also think I wrote a good paper—I can read and write quickly, and English is my first language so these are advantages for me. It is important to attempt every question, and for me, not to go back to the multiple-choice questions as they were very difficult, and if I didn't know the answer the first time, it wasn't going to come to me.

Answer I used homemade flashcards for memorizing lists and concepts for Courses 5, 6, and 8. Courses 1 through 4 were just doing old exam questions over and over again. And I always tried to start studying early (sometimes four to five months in advance).

Answer Mathematics exams (Courses 1 through 4): Do as many practice problems as possible. Classify problems by type for which there is a specific trick to use. Other exams (Course 5 through 8): Read all material quickly (for background). Spend at least 150 to 200 hours learning lists by heart.

Answer I was always studying with a friend. In my case, learning just by reading was quite difficult. Being able to ask questions to a friend and to go through material with someone else was quite helpful. I recommend it a lot to auditory persons, i.e., those who learn more easily by listening than by reading. Also it helps motivation. We often sat in different rooms and got together to review material after couple of hours.

Answer Not to think about how stupid it was to make me learn by heart stuff that I knew I would never use again in my life.

Answer I learn lists by acronyms, usually using the first letter of the first words or the first letter of the most relevant word in the sentence. I also remember the number of elements in a list as well as the elements themselves. This way, you don't waste time trying to find the 6th element when there were five items in the list. Do not focus too much on the first reading—you will retain almost nothing of it. I use it only to make notes in my ACTEX manuals when there is not sufficient information for me to understand what is going on. If you don't understand something at first, skip it. Don't lose too much time. Review the material as many times as you can, going into further details each time.

Answer For the mathematical exams I memorized formulae. For the later exams I read the material through once, then went back over it with the ACTEX study guides and prepared 3×5 index cards with a question on one side and the answer on the other. Once I had been through the entire material again this way I studied exclusively from the index cards, discarding a card once it was memorized. The cards were a convenient size, allowing me to use all my time on the subway for studying as well. I took my study time in two-hour pieces during the working day.

Answer Here is my advice:

▶ *Objectives.* Set objectives based on study material (numbers of papers) instead on numbers of hours.

▶ *Three Readings.* Have at least three readings: the first would be really fast to have a feeling of the material (1 to 2 weeks maximum), the second would be the *real* one where I would read, take notes and do questions and the third one would also be fast but for review only.

▶ *Review Week.* Keep one week to review notes and do past exams. When doing past exams, put yourself in a real situation (3 to 4 hours) to be able to perform similarly at the actual exam.

▶ *Psychology.* Work on your psychological training. As I was saying before, actuarial exams are similar to sportive competitions. You need to visualize the exam, your performance and your success.

Answer Starting the study process at least three months in advance. Cramming doesn't work for these puppies. I would sometimes block off a weekend here or there for fun only and take a break, say after the first reading so as to not feel like I was constantly studying. Going through the material several times—using the study aids, using mnemonic devices such as lists of items to be memorized and making a word or phrase out of the first letter of each item on the list to make it easier to remember, participating in study groups or seminars.

Answer It is always important to remember that you are not writing to achieve a certain grade, but to have a better grade than 60% to 65% of the other people who wrote the exam. Unfortunately, we are now caught in a vicious circle, where a fair proportion of candidates failed the exam on the first try. This means that they have a much better understanding of the exam material the second time around. This makes it that much harder to pass for those writing for the first time. Actuarial students need to sit down and figure out their priorities. Is it family? Is it career? Is it passing exams? Unless you clearly want to pass an exam on first sitting each and every time, don't kill yourself studying. Personally, my study method, which seemed to work for me, was:

▶ *Highlights.* To read through the material once while highlighting the important material (i.e., material which is likely to generate questions). I would read through the highlight a second time while writing a summary (the act of writing it out seemed to help on memorization).

▶ *Exercises.* Next I would start doing exercises (which helps understanding), while reviewing my notes on an occasional basis. I would always skip the most recent three exams' questions.

▶ *Old Exams.* A few weeks before the exam, I would write the exam from three years ago under exam conditions. This is usually a very good wake-up call, to see that one is nowhere near ready. I would do the same one week later, and a few days before the exam.

▶ *Priorities.* If one is dedicated to the exam process, one should not go out on Friday or Saturday nights. That's why it is so important to set up your priorities. I would always do my own summary, never relying on the ACTEX summary since I found them to be, in some instances, inaccurate. When reading through the material, always keep in mind: is this something I would ask a question on? The ability to anticipate the material for question is a great help in focusing the study hours on important subjects.

Answer This is what I would do:

▶ *Two Readings.* Read the material twice.

▶ *Notes.* Take summary notes, including the creation of tables to sort out similar concepts.

▶ *Drill.* Practice with problems a lot.

▶ *Last Month.* For more advanced exams, spend about a month at the end of the study period to learn all my personal notes (In some cases, I had more than 300 pages of notes).

Answer In studying for actuarial exams, what helped me most was planning and discipline. Planning is important because of the quantity of work involved:

▶ *Tables.* I start by creating a table listing all readings from the syllabus, to which I add columns indicating whether I've read the article, typed the notes, worked the problems (twice), and reviewed the article.

▶ *Planning.* Based on the number of articles and pages to read, I determine how many hours I need to study. I usually plan 400 hours per exam. Once the planning stage is completed, discipline is required to stick to the plan.

▶ *Order.* I start by reading the articles in the order they are presented in the syllabus (they are already organized by topic).

▶ *Summaries.* After reading an article, I type notes summarizing important concepts and definitions, and include mathematical examples.

▶ *Practice.* I then work the practice problems from the ACTEX manual.

▶ *Procedure.* I repeat the same procedure for each article. It usually takes two months to read, type the notes and go through the first round of problems.

▶ *One Reading.* Because my notes are very thorough, I read the articles only once.

▶ *Review.* After that, I start reviewing my notes and working the problems for a second time, which takes about a month.

▶ *Memorize.* Two weeks before the exam, I start memorizing lists, definitions, and formulas.

▶ *Sample Exam.* Two days before the exam, I take a few practice exams.

▶ *Timing.* Every day, I record my study time so I can see my progress. At the beginning of each week, I determine how many hours I need to study and how many articles I need to complete to be on schedule. By doing so, what seems to be a mountain of work is broken down into more manageable pieces.

Helpful Study Tools

Actuarial examinations are difficult to pass. The average pass rate is usually below 40%. This means, of course, that the average failure rate is often more than 60%. In other words, many bright and dedicated students who are used to getting high grades in college are having to adjust to the fact that they may actually fail an examination.

To help increase the students' chances of success in actuarial examinations, an extensive commercial support system has developed. Various companies market different types of study tool, organize seminars and special courses, and provide other help for a fee. The survey upon which this book is based suggests that the following study tools are used often used:

▶ The ACTEX study material, produced by ACTEX Publications, Mad River Books and ACTEX Actuarial Recruiting company.

▶ The ASM study material, also marketed by ACTEX.

▶ The JAM study material, also marketed by ACTEX.

▶ The How-To-Pass study manuals, marketed by How-To-Pass.

▶ The Study Aids, marketed by NEAS [New England Actuaries Seminars].

The list is not exhaustive. A search on the Internet will produce additional tools not mentioned by the respondents to the survey. Among these are the study tools made available by CAS on its website. The material included previous examinations, an exam study group, and pass/fail statistics.

Q **Which study aids, such as ACTEX, have you used and which would you recommend? Illustrate your answer with examples related to the SOA and CAS examinations.**

Answer Examinations 1 through 4: How-To-Pass were great! Exam 5: JAM was great! I have found that ACTEX have too much information and not enough explanations. How-To-Pass and JAM really *talk to you*, they are not only summaries of concepts and formulas.

Answer Only ACTEX, as the others didn't exist at the time.

Answer I have actually tried out summaries from all four providers. In my mind, ACTEX is the best all-round, but the ASM books were also really good. I think it is often a good idea to get books from two different companies for a given test, as it is amazing how different the content will be between the various summaries (especially for the higher level tests). The tests are not set by the people who write the study guides, and it is up to the candidates to "guess" what topics and questions will actually be on an SOA test. I have found that most study guides do not adequately cover certain topics, and that using two different study guides offers a certain level of assurance that nothing will be "missed." The textbooks, on the other hand, generally range from mostly useless to completely useless. If time is an issue, and something has to be cut from your study plan, start with the textbooks.

Answer I've used the ACTEX for every exam. It helps, but I often find that they do not explain very well. They take for granted that you know a lot of things. I used it mostly for the exercises. I also use the flashcards. I find them handy because you can always have them on you (in the bus, waiting in line, etc.) and they help me memorize the formulas. I've used the How-To-Pass. I thought they were skipping too much material. They were basically advising us to memorize only a couple formulas. But we don't have time to derive formulas during the exam, and knowing a couple more formulas can make a difference between a 5 and a 6. Finally, I use Study Aids. So far, I like them best. It's clear, they don't explain too fast and there are a lot of exercises. Sometimes, exercises are repetitive, but I remember better by repetition.

Answer For Courses 1 through 3, I used the ACTEX manual, which provides very good summaries, and covers all the topics needed to pass the exam.

Answer For the first four SOA exams, I used the ACTEX manual almost exclusively. Combined with previous exams, I just kept doing practice problems. For Course 5, I used JAM to memorize the material. For Course 6, I also used JAM to memorize, but the actual books were definitely very important for understanding the concepts. I've never used ASM.

Answer The only books I have used were ACTEX manuals.

Answer I only used the ACTEX manuals and really liked them.

Answer For Course 1, I personally felt that the ACTEX manual did not help at all. The questions are too basic and do not illustrate the actual course. The manual is useful if someone does not know the basics, but to prepare for an SOA exam, I did not find it helpful. For the later courses, I have heard the same type of comments about the ACTEX manual.

Answer I used the ACTEX manuals. The other manuals weren't available when I took exams, for the most part. I never went to exam study seminars that cost a ton of money. Students in the United States seem to like them, but I thought they were too expensive for what you got. The new *8I Made Easy* manual, available as a free Internet download, was also helpful.

Answer JAM provides a better summary than ACTEX. For mathematics exams, ACTEX may have more practice problems.

Answer ACTEX is quite good. I do not know the others. I have used the ACTEX manuals for all my exams.

Answer Old exams were my best tools. ACTEX manuals were used a little bit as well.

Answer Using study guides could be good. Do not overemphasize them. The most important thing to really study are old sample SOA exams. The questions come back from year to year. It is possible to learn to answer all the questions without knowing them perfectly. Actually one does not imply the other. In my experience, Exams 1 through 3 have almost no original questions, i.e., questions which have not appeared on a previous exam.

Answer I have always used the ACTEX manuals and they worked for me. Whichever tool you use, I recommended using only one for a given examination. As I remember things visually (by the disposition on the page), seeing it in two or three different ways is more confusing than helpful. Should you be short on time, use JAM (it skips corners). If you want to be more thorough, use ACTEX.

Answer ACTEX. I used them in making index cards. I have never used the others.

Answer The only one that I used was ACTEX. At my time, it was really complete and was really helpful.

Answer I've used the JAM and ASM. I would recommend these over ACTEX, if available because these are usually written by people who take the time to go through the material and make an intelligent comment about the curriculum as opposed to ACTEX which is just a regurgitation of the textbooks with loads and loads of mistakes and typos, and often never peer reviewed before printing. JAM for Course 5 is my favorite. SOA Course 5 is evil, but using the JAM makes it a bit easier. You can order the cue cards. They are excellent for review.

Answer I always used ACTEX to help me with my study. The important word is help. I found that ACTEX is not always accurate, and cannot replace proper reading and understanding of the study material. ACTEX, in my mind, is most useful as a compilation of old exam questions.

Answer I have used ACTEX for SOA exams and Casualty Study Manuals for CAS exams.

Answer The ACTEX manuals provide a good sample of past questions. However, I don't use their summaries. I prefer to type my own notes.

The study material quoted in the survey can be found on the following websites:

1. ACTEX: www.actexmadriver.com
2. ASM: www.studymanuals.com
3. CAS: www.casact.org/admissions/studytools
4. How-To-Pass: www.how-to-pass.com
5. JAM: www.studyjam.com
6. Study Aids: www.neas-seminars.com/misc

Although the survey gives preference to one or two of the six listed study tools, the list of mentioned tools is far from complete. Depending on the search engine used, the Internet query *actuarial study tools*, for example, can produce over 20,000 hits. Since material posted on the Internet is often changed and updated without notice, the mentioned references may have been modified.

2.4 SOA and CAS Course 1

Ideas and Techniques

This course deals with the mathematical foundations of actuarial science. According to the Society of Actuaries, "the purpose of this course is to develop a knowledge of the fundamental mathematical tools for quantitatively assessing risk. The application of these tools to problems encountered in actuarial science is emphasized. A thorough command of calculus and probability topics is assumed. Additionally, a very basic knowledge of insurance and risk management is assumed." (*See* Reference 3, Appendix F, Page 23.)

Why does an actuary need to know calculus? You probably know that some of the basic objects of calculus are the real numbers and differentiable and integrable functions on the real numbers.

Where are real numbers needed in actuarial science? Let us consider a very basic example. Among others, actuarial science deals with the cost of money over time. Interest is the cost of money. You pay interest when you borrow money and you earn interest when you lend money. The amount of interest involved is a function of the amount borrowed, the rate of interest charged, and the length of time for which the money is borrowed. Moreover, in most cases, the interest charged or earned is compound interest. It is calculated over shorter periods than the entire lending period.

Many of the mathematical ideas upon which actuarial science is based are hundreds of years old and have stood the test of time. Calculus dates back to Newton (1642–1727) and Leibniz (1646–1716). History tells us that the theory of probability even predates the beginning of calculus. It is said that a professional gambler named Chevalier de Mere made a great deal of money by betting people that by rolling a die four times he could get at least one six. He was so successful at it that he soon had trouble finding people willing to play his game. So he changed the rules. He started to bet that he could get at least two sixes by rolling a die twenty-four times. Unfortunately for him, he systematically lost. He contacted Pascal (1623–1662) to help explain his losses. Pascal began to correspond with Fermat (1601–1665) to analyze the problem and it is said that thus probability theory was born. Course 1 builds on the ideas and techniques of calculus and probability.

Example 1 *Continuous interest*

Suppose you borrow P dollars at x percent interest for n periods. Using geometric progressions, we can develop for formula for the cost of the loan:

$$P(1+x)^n$$

The formula is based on the assumption that the interest is calculated at the end of each period. However, if the interest is compounded at intervals that is shorter than the interest rate period, then the cost of the loan becomes

$$P\left(1+\frac{x}{t}\right)^{tn}$$

and since

$$(1+x) < \left(1+\frac{x}{2}\right)^2 < \left(1+\frac{x}{3}\right)^3 < \cdots$$

we see that compounding over shorter periods increases the interest paid on the loan. How far can we increase this compounding factor? It turns out that there is a limit beyond which we cannot go. It is called *continuous interest*. Since

$$\lim_{t\to\infty}\left(1+\frac{x}{t}\right)^t = e^x,$$

the largest amount of interest that can be charged for borrowing P dollars at x percent interest for n periods by increasing the compounding intervals to their limit is

$$Pe^{xn}$$

As we can see, even at this very elementary level of finance, the number e, one of the most celebrated of the real numbers, plays a pivotal role. So do limits, the exponential functions e^x, and the idea of continuity. We need calculus to understand these concepts. ▲

It is of course important that these ideas enter the actuarial world at its foundation. Calculus usually remains in the background in day-to-day work of an actuary. Here is what working actuaries and actuarial students have said about their view of the importance of calculus in their work.

According to the SOA, "the purpose of this course is to develop a knowledge of the fundamental mathematical tools for quantitatively assessing risk. The application of these tools to problems encountered in actuarial science is emphasized. A thorough command of calculus and probability topics is assumed. Additionally, a very basic knowledge of insurance and risk management is assumed." (*See* Reference 4, Appendix F, Page 23.)

Examination Topics

The examination consisted of forty multiple-choice questions. They dealt with the following topics from the SOA and CAS syllabus:

Q1   Investment models, exponential functions, logarithmic functions.

Q2   Stocks, dividends, geometric progressions, logarithmic functions.

Q3   Parametric curves, velocity vectors, lengths of vectors, derivatives, cosine functions.

Q4 Random variables, independent random variables, distribution functions, density functions, expected values, maximum-value functions, probabilities, sine and cosine functions, integrals.

Q5 Functions defined by cases, property and casualty insurance, density functions, joint density functions, probabilities, double integrals.

Q6 Life insurance (standard, preferred, and ultra-preferred), probabilities, conditional probabilities, Bayes' formula.

Q7 Functions defined by cases, joint density functions, covariance, double integrals, expected values.

Q8 Capital, labor, production rates of change, chain rule.

Q9 Health insurance, risk factors, probabilities, unconditional probabilities, algebra of sets.

Q10 Life insurance, premiums, survival functions, expected values, probabilities.

Q11 Insurance products, functional models, derivatives, maximum-value test.

Q12 Probabilities, algebra of sets.

Q13 Health insurance, probabilities, algebra of sets, independence of events.

Q14 Functions defined by cases, stock prices modeled with random variables, joint density distributions, conditional variance, marginal density functions, integrals, expected values.

Q15 Parametric curves, slopes, tangents, derivatives.

Q16 Functions defined by cases, ratios, graphs, concavity, step functions.

Q17 Functions defined by cases, exponential functions, auto insurance, probability distributions, probability density functions, expected values, integrals, maximum-value test.

Q18 Exponential functions, partial derivatives, rates of change.

Q19 Lifetimes, independent lifetimes, means, variances, normal distributions, random variables, probabilities, minimum values.

Q20 Continuous measurements, time-to-failure, exponential distributions, means, expected values, integrals, maximum values.

Q21 Differential equation models for diseases, separable differential equations, general solutions, partial fractions, integrals.

Q22 Auto insurance, insurance claims, random variables, exponential distributions, means, probabilities, algebra of sets.

Q23 Quality control, probabilities, Bayes' formula.

Q24 Lifetimes, joint density functions, probabilities, double integrals.

Q25 Volumes, surface areas, spheres, rates of change, derivatives.

Q26 Earthquake insurance, premiums modeled by random variables, exponential random variables, independent random variables, probability density functions, means, double integrals, derivatives.

Q27 Functions defined by cases, property and casualty insurance, damage claims, independent random variables, joint density functions, expected values, derivatives, improper integrals.

Q28 Functions defined by cases, mortality functions integrals, inequalities, minimum values.

Q29 Insurance risk modeled by random variables, probabilities, trinomials, probability functions.

Q30 Profit models, fixed costs, variable costs, maximum profit, cost functions, revenue functions, profit functions, quadratic functions.

Q31 Group health insurance, supplementary coverage, probabilities, Venn diagrams.

Q32 Time-to-failure, exponential distributions, means, variances.

Q33 Insurance claims, normal distribution of claims, means, standard deviations, independent random variables, expected values, linear combinations.

Q34 Graphs of functions, graphs of derivatives, slopes.

Q35 Property and casualty insurance, time-to-failure, density functions, variances, expected values, integrals.

Q36 Pollution models, averages, double integrals.

Q37 Loss models, probabilities, expected values.

Q38 Functions defined by cases, continuity at a point.

Q39 Functions defined by cases, home insurance, random variables, probability distributions, density functions, integrals.

Q40 Life insurance, probabilities, conditional probabilities, algebra of sets.

If we examine the frequencies of some of the topics and techniques tested in this examination, we come up with the following result: probability (30/40), functions (20/40), integrals (15/40), random variables (14/40), derivatives (10/40), expected values (10/40).

Questions and Answers

Here are some examples of how these ideas and techniques were tested.

Question 4 *A company agrees to accept the highest of four sealed bids on a property. The four bids are regarded as four independent random variables with common cumulative distribution function*

$$F(x) = \frac{1}{2}(1 + \sin \pi x)$$

for $\frac{3}{2} \le x \le \frac{5}{2}$. *What is the expected value of the accepted bid?*

Answer Let X_1, X_2, X_3, and X_4 denote the four independent bids with common distribution function F. Then if we define $Y = \max(X_1, X_2, X_3, X_4)$, the distribution function G of Y is given by

$$\begin{aligned} G(y) &= \Pr[Y \le y] \\ &= \Pr[(X_1 \le y) \cap (X_2 \le y) \cap (X_3 \le y) \cap (X_4 \le y)] \\ &= \Pr[X_1 \le y]\Pr[X_2 \le y]\Pr[X_3 \le y]\Pr[X_4 \le y] \\ &= [F(y)]^4 \\ &= \frac{1}{16}(1 + \sin \pi y)^4 , \frac{3}{2} \le y \le \frac{5}{2} \end{aligned}$$

It follows that the density function g of Y is given by

$$\begin{aligned} g(y) &= G'(y) \\ &= \frac{1}{4}(1 + \sin \pi y)^3 (\pi \cos \pi y) \\ &= \frac{\pi}{4} \cos \pi y (1 + \sin \pi y)^3 , \frac{3}{2} \le y \le \frac{5}{2} \end{aligned}$$

Therefore,

$$\begin{aligned} E[Y] &= \int_{3/2}^{5/2} y g(y)\, dy \\ &= \int_{3/2}^{5/2} \frac{\pi}{4} y \cos \pi y (1 + \sin \pi y)^3\, dy \\ &\approx 2.22656 \end{aligned}$$

This answers the question. ▲

Question 10 *Two life insurance policies, each with a death benefit of 10,000 and a one-time premium of 500, are sold to a couple, one for each person. The policies will expire at the end of the tenth year. The probability that only the wife will survive at least ten years is 0.025, the probability that only the husband will survive at least ten years is 0.01, and the probability that both of them will survive at least ten years is 0.96. What is the expected excess of premiums over claims, given that the husband survives at least ten years?*

Answer Let W be the event that the wife survives at least 10 years, H the event that the husband survives at least 10 years, B the paid benefit, and P the profit from selling the policies. Then

$$\Pr[H] = Pr[H \cap W] + Pr[H \cap W^c] = 0.96 + 0.01 = 0.97$$

and

$$\Pr[W^c \mid H] = \frac{\Pr[H \cap W^c]}{\Pr[H]} = \frac{0.01}{0.97} = 0.0103$$

It follows that

$$\begin{aligned}
E[P] &= E[1000 - B] \\
&= 1000 - E[B] \\
&= 1000 - \{(0)\Pr[W \mid H] + (10,000)\Pr[W^c \mid H]\} \\
&= 1000 - 10,000(0.0103) \\
&= 1000 - 103 \\
&= 897
\end{aligned}$$

This answers the question. ▲

Question 11 *An insurance company has* $160,000$ *to spend on the development and marketing of a new insurance policy. If x is spent on development and y is spent on marketing, then*

$$\frac{x^{1/4}y^{3/4}}{1000}$$

policies will be sold during the first year. Calculate the maximum possible number of policies the company can sell during the first year.

Answer Observe that x and y follow the constraint equation

$$\begin{aligned}
x + y &= 160,000 \\
x &= 160,000 - y, \text{ where } 0 \le y \le 160,000
\end{aligned}$$

Using this constraint equation, we express the policy sales $g(x,y)$ as a function $f(y)$ of marketing y:

$$f(y) = g(160,000 - y, y) = 0.001(160,000 - y)^{1/4}y^{3/4}$$

We then compute $f'(y)$:

$$\begin{aligned}
f'(y) &= \left\{-\frac{1}{4}(160,000 - y)^{-3/4}y^{3/4} + \frac{3}{4}(160,000 - y)^{1/4}y^{-1/4}\right\}/1000 \\
&= -\frac{1}{4000}(160,000 - y)^{-3/4}y^{-1/4}[y - 3(160,000 - y)]
\end{aligned}$$

$$= -\frac{1}{4000}(160,000-y)^{-3/4}y^{-1/4}(4y-480,000)$$
$$= \frac{1}{1000}(160,000-y)^{-3/4}y^{-1/4}(120,000-y), \; 0 \le y \le 160,000$$

and note that
$$f'(y) > 0 \text{ for } 0 \le y < 120,000,$$
$$f'(y) = 0 \text{ for } y = 120,000, \text{ and}$$
$$f'(y) < 0 \text{ for } 120,000 < y < 160,000$$

Therefore sales are maximized when $y = 120,000$. It follows that
$$f(120,000) = 0.001(160,000 - 120,000)^{1/4}(120,000)^{3/4} = 91.2$$

maximizes f.▲

Question 13 *A study is being conducted in which the health of two independent groups of ten policyholders is being monitored over a one-year period of time. Individual participants in the study drop out before the end of the study with probability 0.2 (independently of the other participants). What is the probability that at least 9 participants complete the study in one of the two groups, but not in both groups?*

Answer Let X be the number of group 1 participants that complete the study, and let Y be the number of group 2 participants that complete the study. Since
$$P[X \ge 9] = \left[\binom{10}{9}(0.2)(0.8)^9 + \binom{10}{10}(0.8)^{10} \right] = 0.376,$$

it follows that
$$P\{[(X \ge 9) \cap (Y < 9)] \cup [(X < 9) \cap (Y \ge 9)]\}$$
$$= P[(X \ge 9) \cap (Y < 9)] + P[(X < 9) \cap (Y \ge 9)]$$
$$= 2P[(X \ge 9) \cap (Y < 9)] \text{ (due to symmetry)}$$
$$= 2P[X \ge 9]P[Y < 9] \text{ (due to independence)}$$
$$= 2P[X \ge 9]P[X < 9] \text{ (again due to symmetry)}$$
$$= 2P[X \ge 9](1 - P[X \ge 9])$$
$$= 2[0.376][1 - 0.376] = 0.469.$$

This answers the question. ▲

Question 14 *The stock prices of two companies at the end of any given year are modeled with random variables X and Y that follow a distribution with joint density function*

$$f(x,y) = \begin{cases} 2x & for\ 0 < x < 1, x < y < x+1 \\ 0 & otherwise \end{cases}$$

What is the conditional variance of Y given that X = x?

Answer Let $f_1(x)$ denote the marginal density function of X. Then

$$f_1(x) = \int_x^{x+1} 2x\,dy = 2xy|_x^{x+1} = 2x(x+1-x) = 2x,\ 0 < x < 1$$

Consequently,

$$f(y \mid x) = \frac{f(x,y)}{f_1(x)} = \begin{cases} 1 & if\ x < y < x+1 \\ 0 & otherwise \end{cases}$$

and

$$E[Y \mid X] = \int_x^{x+1} y\,dy = \frac{1}{2}y^2\Big|_x^{x+1} = \frac{1}{2}(x+1)^2 - \frac{1}{2}x^2$$

$$= \frac{1}{2}x^2 + x + \frac{1}{2} - \frac{1}{2}x^2 = x + \frac{1}{2}$$

$$E[Y^2 \mid X] = \int_x^{x+1} y^2\,dy = \frac{1}{3}y^3\Big|_x^{x+1} = \frac{1}{3}(x+1)^3 - \frac{1}{3}x^3$$

$$= \frac{1}{3}x^3 + x^2 + x + \frac{1}{3} - \frac{1}{3}x^3 = x^2 + x + \frac{1}{3}$$

$$Var[Y \mid X] = E[Y^2 \mid X] - \{E[Y \mid X]\}^2 = x^2 + x + \frac{1}{3} - \left(x + \frac{1}{2}\right)^2$$

$$= x^2 + x + \frac{1}{3} - x^2 - x - \frac{1}{4} = \frac{1}{12}$$

This answers the question. ▲

Question 17 *An auto insurance company insures an automobile worth 15,000 for one year under a policy with a 1,000 deductible. During the policy year there is a 0.04 chance of partial damage to the car and a 0.02 chance of a total loss of the car. If there is partial damage to the car, the amount X of damage (in thousands) follows a distribution with density function*

$$f(x) = \begin{cases} 0.5003e^{-x/2} & for\ 0 < x < 15 \\ 0 & otherwise \end{cases}$$

What is the expected claim payment?

Answer Let Y denote the claim payment made by the insurance company. Then

$$Y = \begin{cases} 0 & \text{with probability } 0.94 \\ \max(0, X - 1) & \text{with probability } 0.04 \\ 14 & \text{with probability } 0.02 \end{cases}$$

and

$$E[Y] = (0.94)(0) + (0.04)(0.5003) \int_1^{15} (x-1) e^{-x/2} dx + (0.02)(14)$$

$$= (0.020012) \left[\int_1^{15} x e^{-x/2} dx - \int_1^{15} e^{-x/2} dx \right] + 0.28$$

$$= 0.28 + (0.020012) \left[-2x e^{-x/2} \Big|_1^{15} + 2 \int_1^{15} e^{-x/2} dx - \int_1^{15} e^{-x/2} dx \right]$$

$$= 0.28 + (0.020012) \left[-30 e^{-7.5} + 2 e^{-0.5} + \int_1^{15} e^{-x/2} dx \right]$$

$$= 0.28 + (0.020012) \left(-30 e^{-7.5} + 2 e^{-0.5} - 2 e^{-7.5} + 2 e^{-0.5} \right)$$

$$= 0.28 + (0.020012) \left(-32 e^{-7.5} + 4 e^{-0.5} \right)$$

$$= 0.28 + (0.020012)(2.408)$$

$$= 0.328 \text{ (in thousands)}$$

It follows that the expected claim payment is 328. ▲

Question 19 *A company manufactures light bulbs with a lifetime, in months, that is normally distributed with mean 3 and variance 1. A consumer buys a number of these bulbs with the intention of replacing them successively as they burn out. The light bulbs have independent lifetimes. What is the smallest number of bulbs to be purchased so that the succession of light bulbs produces light for at least 40 months with probability at least 0.9772?*

Answer Let $X_1, \ldots, X_n$ denote the life spans of the n light bulbs purchased. Since these random variables are independent and normally distributed with mean 3 and variance 1, the random variable

$$S = X_1 + \cdots + X_n$$

is also normally distributed with mean $\mu = 3n$ and standard deviation $\sigma = \sqrt{n}$. We want to choose the smallest value for n such that

$$0.9772 \leq \Pr\left[S > 40\right] = \Pr\left[\frac{S - 3n}{\sqrt{n}} > \frac{40 - 3n}{\sqrt{n}}\right]$$

Recalling that in the case of a normal distribution, the probability that an observation is above two standard deviations below the mean is 0.9772, we conclude that n should satisfy the following inequality:

$$-2 \geq \frac{40 - 3n}{\sqrt{n}}$$

To find such an n, we solve the corresponding equation for n :

$$-2 = \frac{40 - 3n}{\sqrt{n}}$$
$$-2\sqrt{n} = 40 - 3n$$
$$3n - 2\sqrt{n} - 40 = 0$$
$$\left(3\sqrt{n} + 10\right)\left(\sqrt{n} - 4\right) = 0$$
$$\sqrt{n} = 4$$
$$n = 16$$

This answers the question. ▲

Question 21 *The rate at which a disease spreads through a town can be modeled by the differential equation*

$$\frac{dQ}{dt} = Q\left(N - Q\right)$$

where $Q(t)$ is the number of residents infected at time t and N is the total number of residents. Find $Q(t)$.

Answer The given equation has the obvious singular solutions $Q(t) \equiv 0$ and $Q(t) \equiv N$, but these are of little relevance to the given problems. To find the other solutions, we proceed as follows. The differential equation that we are given is separable. As a result, the general solution is given by

$$\int \frac{1}{Q\left(N - Q\right)} dQ = \int dt = t + C$$

where C is a constant. To calculate the integral on the left-hand side of this equation, we determine the partial fractions of the integrand. In other words, we need to find constants A and B such that

$$\frac{1}{Q(N-Q)} = \frac{A}{Q} + \frac{B}{N-Q}$$
$$1 = A(N-Q) + BQ$$
$$1 = AN + (B-A)Q$$

Therefore, $AN = 1$ and $B - A = 0$. Hence $B = A = 1/N$ and

$$\int \frac{1}{Q(N-Q)} dQ = \frac{1}{N} \int \frac{1}{Q} + \frac{1}{N} \int \frac{1}{N-Q} dQ$$
$$= \frac{1}{N} \ln Q - \frac{1}{N} \ln(N-Q) + K$$
$$= \frac{1}{N} \ln \left[\frac{Q}{N-Q} \right] + K,$$

where K is a constant.
Consequently,

$$\frac{1}{N} \ln \left[\frac{Q}{N-Q} \right] + K = t + C$$
$$\left(\frac{Q}{N-Q} \right)^{1/N} e^K = e^t e^C$$
$$\left(\frac{Q}{N-Q} \right)^{1/N} = e^t e^{C-K}$$
$$\frac{Q}{N-Q} = e^{Nt} e^{N(C-K)},$$

so that

$$Q = ae^{Nt}(N-Q)$$
$$= aNe^{Nt} - ae^{Nt}Q, \text{where } a = e^{N(C-K)} \text{ is a constant}$$
$$(1 + ae^{Nt})Q = aNe^{Nt}$$
$$Q(t) = \frac{aNe^{Nt}}{1 + ae^{Nt}}$$

This answers the question. ▲

Question 26 *A company offers earthquake insurance. Annual premiums are modeled by an exponential random variable with mean 2. Annual claims are modeled by an exponential random variable with mean 1. Premiums and claims are* *independent. Let X denote the ratio of claims to premiums. What is the density function of X ?*

Answer Let U be the annual claims, V the annual premiums, $g(u, v)$ the joint density function of U and V, $f(x)$ the density function of X, and $F(x)$ the distribution function of X. Then, since U and V are independent,

$$g(u, v) = \left(e^{-u}\right)\left(\frac{1}{2}e^{-v/2}\right) = \frac{1}{2}e^{-u}e^{-v/2}, \ 0 < u < \infty, \ 0 < v < \infty$$

and

$$F(x) = \Pr\left[\frac{U}{V} \leq x\right] = \Pr[U \leq Vx]$$

$$= \int_0^\infty \int_0^{vx} g(u, v)\, du\, dv = \int_0^\infty \int_0^{vx} e^{-u}e^{-v/2}\, du\, dv$$

$$= \int_0^\infty -\frac{1}{2}e^{-u}e^{-v/2}\Big|_0^{vx}\, dv = \int_0^\infty \left(-\frac{1}{2}e^{-vx}e^{-v/2} + \frac{1}{2}e^{-v/2}\right)\, dv$$

$$= \int_0^\infty \left(-\frac{1}{2}e^{-v(x+1/2)} + \frac{1}{2}e^{-v/2}\right)\, dv$$

$$= \left[\frac{1}{2x+1}e^{-v(x+1/2)} - e^{-v/2}\right]_0^\infty$$

$$= -\frac{1}{2x+1} + 1$$

It follows that

$$f(x) = F'(x) = \frac{2}{(2x+1)^2}$$

This answers the question. ▲

Question 27 *Claim amounts for wind damage to insured homes are independent random variables with common density function*

$$f(x) = \begin{cases} \frac{3}{x^4} & \text{for } x > 1 \\ 0 & \text{otherwise} \end{cases}$$

where x is the amount of a claim in thousands. Suppose 3 such claims will be made. What is the expected value of the largest of the three claims?

Answer First, observe that the distribution function of X is given by

$$F(x) = \int_1^x \frac{3}{t^4}\, dt = -\frac{1}{t^3}\Big|_1^x = 1 - \frac{1}{x^3}, \ x > 1$$

Next, let $X_1, X_2,$ and X_3 denote the three claims made that have this distribution. Then if Y denotes the largest of these three claims, it follows that the distribution function of Y is given by

$$G(y) = \Pr[X_1 \le y] \Pr[X_2 \le y] \Pr[X_3 \le y]$$
$$= \left(1 - \frac{1}{y^3}\right)^3, \ y > 1$$

while the density function of Y is given by

$$g(y) = G'(y) = 3\left(1 - \frac{1}{y^3}\right)^2 \left(\frac{3}{y^4}\right) = \left(\frac{9}{y^4}\right)\left(1 - \frac{1}{y^3}\right)^2, \ y > 1$$

Therefore,

$$E[Y] = \int_1^\infty \frac{9}{y^3}\left(1 - \frac{1}{y^3}\right)^2 dy = \int_1^\infty \frac{9}{y^3}\left(1 - \frac{2}{y^3} + \frac{1}{y^6}\right) dy$$
$$= \int_1^\infty \left(\frac{9}{y^3} - \frac{18}{y^6} + \frac{9}{y^9}\right) dy = \left. -\frac{9}{2y^2} + \frac{18}{5y^5} - \frac{9}{8y^8}\right|_1^\infty$$
$$= 9\left[\frac{1}{2} - \frac{2}{5} + \frac{1}{8}\right] = 2.025 \text{ (in thousands)}$$

This answers the question. ▲

Question 29 *A large pool of adults earning their first driver's license includes 50% low-risk drivers, 30% moderate-risk drivers, and 20% high-risk drivers. Because these drivers have no prior driving record, an insurance company considers each driver to be randomly selected from the pool. This month, the insurance company writes 4 new policies for adults earning their first driver's license. What is the probability that these 4 will contain at least two more high-risk drivers than low-risk drivers?*

Answer Let X be the number of low-risk drivers insured, Y the number of moderate-risk drivers insured, Z the number of high-risk drivers insured, and $f(x,y,z)$ the probability function of X, Y, and Z. Then f is a trinomial probability function, so

$$\Pr[z \ge x+2] = f(0,0,4) + f(1,0,3) + f(0,1,3) + f(0,2,2)$$
$$= (0.20)^4 + 4(0.50)(0.20)^3$$
$$+ 4(0.30)(0.20)^3 + \frac{4!}{2!2!}(0.30)^2(0.20)^2$$
$$= 0.0488$$

This answers the question. ▲

Question 36 *A town in the shape of a square with each side measuring* 4 *has an industrial plant at its center. The industrial plant is polluting the air such that the concentration of pollutants at each location* (x,y) *in the town can be modeled by the function*

$$C(x,y) = 22,500\left(8 - x^2 - y^2\right) \text{ for } -2 \le x \le 2 \text{ and } -2 \le y \le 2.$$

Calculate the average pollution concentration over the entire town.

Answer Let T denote the total concentration of pollutants over the town. By symmetry we have

$$T = 4 \int_0^2 \int_0^2 22,500\left(8 - x^2 - y^2\right) dxdy$$

$$= (4)(7500) \int_0^2 \left(24x - x^3 - 3xy^2\right)\big|_0^2 dy$$

$$= 30,000 \int_0^2 \left(48 - 8 - 6y^2\right) dy$$

$$= 30,000 \int_0^2 \left(40 - 6y^2\right) dy$$

$$= 30,000 \left(40y - 2y^3\right)\big|_0^2 = 30,000\left(80 - 16\right)$$

$$= 30,000\left(64\right) = 1,920,000$$

Since the town covers 16 square miles, it follows that the average pollution concentration A is

$$A = T/16 = 1,920,000/16 = 120,000$$

This answers the question. ▲

Question 37 *A tour operator has a bus that can accommodate* 20 *tourists. The operator knows that tourists may not show up, so he sells* 21 *tickets. The probability that an individual tourist will not show up is* 0.02, *independent of all other tourists. Each ticket costs* 50, *and is non-refundable if a tourist fails to show up. If a tourist shows up and a seat is not available, the tour operator has to pay* 100 (*ticket cost* + 50 *penalty*) *to the tourist. What is the expected revenue of the tour operator?*

Answer Observe that the bus driver collects $21 \times 50 = 1050$ for the 21 tickets he sells. However, he may be required to refund 100 to one passenger if all 21 ticket holders show up. Since passengers show up or do not show up independently of one another, the probability that all 21 passengers will show up is

$$(1 - 0.02)^{21} = (0.98)^{21} = 0.65$$

Therefore, the tour operator's expected revenue is $1050 - (100)(0.65) = 985$. ▲

Question 39 *An insurance company insures a large number of homes. The insured value X of a randomly selected home is assumed to follow a distribution with density function*

$$f(x) = \begin{cases} 3x^{-4} & \text{for } x > 1 \\ 0 & \text{otherwise} \end{cases}$$

Given that a randomly selected home is insured for at least 1.5, what is the probability that it is insured for less than 2?

Answer Let F denote the distribution function of f. Then

$$F(x) = \Pr[X \le x] = \int_1^x 3t^{-4}dt = -t^{-3}\Big|_1^x = 1 - x^{-3}$$

Therefore,

$$\Pr[X < 2 \mid X \ge 1.5] = \frac{\Pr[(X < 2) \cap (X \ge 1.5)]}{\Pr[X \ge 1.5]}$$

$$= \frac{\Pr[X < 2] - \Pr[X \le 1.5]}{\Pr[X \ge 1.5]}$$

$$= \frac{F(2) - F(1.5)}{1 - F(1.5)} = \frac{(1.5)^{-3} - (2)^{-3}}{(1.5)^{-3}}$$

$$= 1 - \left(\frac{3}{4}\right)^3 = 0.578$$

This answers the question. ▲

Question 40 *A public health researcher examines the medical records of a group of 937 men who died in 1999 and discovers that 210 of the men died from causes related to heart disease. Moreover, 312 of the 937 men had at least one parent who suffered from heart disease, and, of these 312 men, 102 died from causes related to heart disease. Determine the probability that a man randomly selected from this group died of causes related to heart disease, given that neither of his parents suffered from heart disease.*

Answer Let H be the event that a death is due to heart disease, and F be the event that at least one parent suffered from heart disease.

Based on the medical records, we have

$$P[H \cap F^c] = \frac{210 - 102}{937} = \frac{108}{937}$$

$$P[F^c] = \frac{937 - 312}{937} = \frac{625}{937}$$

and

$$P[H \mid F^c] = \frac{P[H \cap F^c]}{P[F^c]} = \frac{108}{937} \div \frac{625}{937} = \frac{108}{625} = 0.173$$

This answers the question. ▲

2.5 SOA and CAS Course 2

Ideas and Techniques

This course deals with four related fields of knowledge: microeconomics, macroeconomics, finance, and the theory of interest.

Microeconomics

Microeconomics focuses on the role of individual firms and groups of firms with national and international economies. Key ideas of microeconomics are the demand and supply for individual goods and services, their trading and patterns of pricing, market equilibrium, and idea such as concepts as monopoly, where one firm dominates the market, and oligopoly, where a small number of firms dominate a national or global market. According to the SOA syllabus, actuaries should "be able to use the following microeconomic principles to build models to increase their understanding of the framework of contingent events and to use as a frame for activities such as pricing," and "be able to use knowledge of the following microeconomic principles to increase their understanding of the markets in which we operate and of the regulatory issues."

Macroeconomics

Macroeconomics deals with aggregate economic factors such as total national income and output, employment, balance of payments, rates of inflation, and the business cycle. One of the key ideas of macroeconomics is that of a gross national product: the total value of goods and services produced in an economy during a specified period time. According to the SOA syllabus, actuaries should understand macroeconomic principles to be able to develop economic models and assess the consequences of macroeconomics assumptions. They should understand "the relationship among interest rates, demand for money, consumption and investment using concepts such as the IS/LM (IS: investment = savings, LM: demand for money = supply of money) curve, fiscal and monetary policy, and how foreign exchange rates affect the gross national product and national income." They should understand macroeconomic principles and know how to relate them to the business cycle.

Theory of Interest

The theory of interest is at the heart of actuarial science. It deals with the cost of money over time. According to the SOA syllabus, actuaries should understand how the theory of interest is used in annuity functions and be able to "apply the concepts of present and accumulated value for various streams of cash flows as a basis for future use in: reserving, valuation, pricing, duration, asset/liability management, investment income, capital budgeting, and contingencies." Calculus plays a major role in the theory of interest since exponential functions infinite series, and the continuous measurement of interest are key elements of financial modeling. Actuaries also need to be able to determine "the yield rates on investments and the time required to accumulate a given amount or repay a given loan amount," and use annuity functions in financial context such as mortgages and similar products.

The starting point of actuarial science is that of an *annuity*, the idea of investing money, earning interest on the investment, and receiving payments in return. If you invest $1,000 at 5% interest per year, the bank will pay you an annuity of $50 per year to you or your heirs or until you withdraw your initial investment.

A fundamental variation on this theme is the idea of a *life annuity*. If you pay $1,000 to a life insurance company, the company may contract to pay you a fixed amount until you die. At that point the payments will cease and the initial investment is not refunded. The amount the company agrees to pay depends both on prevailing rates of interest and on how long the company expects you to live. Retirement benefits from pension plans are typical life annuities. This is the point where probability and statistics enter the picture. Most countries collect statistics on life expectancy and update this information on a period basis. The results are *life tables*. The so-called *Breslau Table* seems to be the first published record of this kind. In 1693, Edmond Halley published his analysis of the records of death of the city of Breslau in Germany (now Wroclaw, Poland). He started with a population of 1000 aged one, and calculated the number of survivors at different ages, up to the age of 84. Based on his table, Halley developed a method for calculating the premiums of life annuities dependent on two lives. One of the difficulties of using Halley's table for the purpose of calculating life annuities premiums was that the numbers in the table do not give rise to an obvious formula for mortality. A few years after Halley published his table, de Moivre tried to remedy this situation by postulating that the number of survivors decreased in arithmetical progression. If N is the initial population in any given year and d is the number of deaths per year, then the number of survivors k years later is $N - kd$. It turned out that de Moivre's assumption produced results that were sufficiently close to those of Halley to be of practical value.

Here are some examples of basic *annuity functions* essential for actuarial work.

Example 1 *Accumulated value of an annuity*

The function

$$S = Rs_{\overline{n}|i} = R\frac{(1+i)^n - 1}{i}$$

calculates the accumulated value S of an ordinary simple annuity of n payments of R dollars per payment. The expression

$$s_{\overline{n}|i} = \frac{(1+i)^n - 1}{i}$$

is known as the "accumulation factor for n payments," and is read as "s angle n at i. " ▲

Example 2 *Discounted value of an annuity*

The function

$$A = Ra_{\overline{n}|i} = R\frac{1 - (1+i)^{-n}}{i}$$

calculates the present value of the set of payments R due one period before the first payment. The expression

$$a_{\overline{n}|i} = \frac{1 - (1+i)^{-n}}{i}$$

is known as the present value of an annuity-immediate of a payment of one for n periods, i.e., when the payment is made at the *end* of each period. ▲

Example 3 *Present value of an annnuity-due*

The expression

$$\ddot{a}_{\overline{n}|i} = (1+i)a_{\overline{n}|i}$$

is known as the present value of an annuity-due of a payment of one for n periods, i.e., when the payment is made that the *beginning* of each period. ▲

Example 4 *Accumulated value of an annuity-due*

The expression

$$\ddot{s}_{\overline{n}|i} = (1+i)s_{\overline{n}|i}$$

denotes the accumulated value of an annuity-due of one at time n. ▲

Here are some typical life functions used by actuaries.

Example 5 *Probability of death*

The expression

$$q_x$$

denotes the probability that an individual, alive at age x, will die before age $x + 1$. ▲

Example 6 *Probability of survival*

The expression

$$p_x = 1 - q_x$$

denotes the probability that an individual, alive at age x, will survive beyond age $x + 1$. ▲

Example 7 *Bounded probability of death*

The expression

$$_nq_x$$

denotes the probability that an individual, alive at age x, will die before age $x + n$. ▲

Example 8 *Bounded probability of survival*

The expression

$$_np_x = 1 - _nq_x$$

denotes the probability that an individual, alive at age x, will survive beyond age $x + n$. ▲

The death and survival probabilities are used to define basic life insurance products. Here are some examples:

Example 9 *Pure endowment*

The function

$$_nE_x = (1 + i)^{-n} {}_np_x$$

computes the cost of an n-year endowment of one dollar to be paid to a person aged x years if that person reaches age $x + n$. ▲

Example 10 *Discounted value*

The function

$$a_x = \sum_{t=1}^{\infty} (1 + i)^{-t} {}_tp_x$$

computes the discounted value of a one-dollar ordinary life annuity issued to someone of age x. ▲

Example 11 *Life annuity value*

The function

$$\ddot{a}_x = 1 + \sum_{t=1}^{\infty} (1+i)^{-t} \, {}_tp_x$$

computes the value of a one-dollar life annuity issued to someone of age x whose first premium payment is due now. ▲

Example 12 *Discounted value of a life annuity due*

The function

$$a_{x:\overline{n}|} = \sum_{t=1}^{n} (1+i)^{-t} \, {}_tp_x$$

computes the discounted value of a one-dollar temporary life annuity due when n is the length of the payment period. ▲

Example 13 *Value of a life annuity due*

The function

$$\ddot{a}_{x:\overline{n}|} = \sum_{t=0}^{n-1} (1+i)^{-t} \, {}_tp_x$$

computes the value of a one-dollar life annuity due issued to someone of age x whose first premium payment is due now. ▲

Example 14 *Net single premium of term insurance*

The function

$$A^1_{x:\overline{n}|} = \sum_{t=0}^{n-1} (1+i)^{-(t+1)} \, {}_tp_x q_{x+t}$$

computes the net single premium for a one-dollar, n-year term insurance policy sold to a person x years old. ▲

Example 15 *Net single premium of whole life insurance*

The function

$$A_x = \sum_{t=0}^{\infty} (1+i)^{-(t+1)} \, {}_tp_x q_{x+t}$$

computes the net single premium for a one-dollar whole life insurance policy sold to a person x years old. ▲

Example 16 *Net single premium of an endowment*

The function

$$A^1_{x:\overline{n}|} + {}_n E_x$$

computes the net single premium for one-dollar, n-year endowment insurance policy. ▲

Example 17 *Annual premium of a whole life insurance*

The function

$$P_x = \frac{A_x}{\ddot{a}_x}$$

computes the net annual premium for a one-dollar whole life insurance policy. ▲

The finance part of Course 2 deals with financial statements including balance sheets, income statements, and statements of cash flow. The main ideas involved are discounted cash flow, internal rate of return, present and future values of bonds and apply the dividend growth model and price/earnings ratios concept to valuing stocks. Actuaries must be able to assess financial performance using "net present value and the payback, discounted payback models, internal rate of return and profitability index models."Among the key ideas are risk and return, and efficient markets. Actuaries must be able to valuate securities, and apply measures of portfolio risk, analyze the effects of diversification, systematic and unsystematic risks. They must be able to calculate portfolio risks and analyze the impact of individual securities on portfolio risks and identify efficient portfolios and apply the CAPM [capital asset pricing model] to measure the cost of capital. They must also understand "the impact of financial leverage and long- and short-term financing policies on capital structure, sources of capital and the definitions of techniques for valuing basic options such as calls and puts."

Examination Topics

The 2001 examination consisted of 50 multiple-choice questions. They dealt with the following topics from the SOA and CAS syllabus:

Q1 Efficient market hypothesis, market prices, past price data, actively managed portfolios, semi-strong version of the efficient market theory, securities.

Q2 Free rider problem, public goods, private markets, governments, non-paying consumers, social costs, positive prices, long-run marginal costs, long-run average costs.

Q3 Economies, contractionary phases, business cycles, indicators, downturn, unemployment rates, building permits, stock prices, delivery lags, business inventories, inventory accumulation.

Q4   Loans, amortizations, annual payments, effective rates, sinking funds.

Q5   Perpetuity-immediate annuities, present value.

Q6   Constant-cost competitive industries, long-run equilibrium, licensing fees, long-run market supply, long-run firm supply, fixed costs, prices, demand, average costs, marginal costs, output.

Q7   Loan, nominal interest rates, compound interest, lump sums, interest.

Q8   Shares, common stock, capital, earnings, treasury stock.

Q9   Consumer goods, marketplace, demand, prices, Engel curve, demand curve, personal income, income effect, substitution effect, compensated demand curve, normal goods, uncompensated demand curve, income elasticity, slopes, compensated price decline.

Q10 Investment, return level cash flow, internal rates of return, single cash flow, risk-free rates, market risk premiums, estimated beta, payback periods, net present value, annual cash flow, annuities, discount rates.

Q11 Salvage value, depreciation, declining balance method, sum-of-the-years digits method.

Q12 Investments, annual effective discount rates, interest.

Q13 Loans, present value, interest, principal.

Q14 Productivity, productivity growth, government spending, infrastructure, capital stock, real wage growth, workforce, demographic composition, labor, capital, economy, service jobs, real wages, real wage growth.

Q15 Production costs, delivery costs, equilibrium price, largest daily rate, outsourcing, profits.

Q16 Money supply, central bank, commercial banking system, public demand for currency, market interest rate, exogenous increase in interest rates, discount rate, reserve requirement ratio, bond sale, reserves.

Q17 Effective rates of interest, present value, perpetuity, present value, perpetuity-immediate.

Q18 Natural monopolies, prices, marginal costs, loss, fixed costs, marginal cost curves, industry demand curves, marginal revenue, average cost curves, competitive prices.

Q19 Net cash flow, opportunity costs, capital, net present value, expected economic income, cash flow.

Q20 Long-run real output, velocity of money, growth, monetary authority, target rates of inflation, money supply, price levels, growth rates.

Q21 Demand, supply, elasticity, exogenous increase in wages, prices, quantities, marginal production costs, equilibrium prices.

Q22 Investment, cash flow, after-tax weighted average costs of capital, net present value, equity financing, debt financing, marginal tax rates, equity costs, debt costs.

Q23 Real income, interest rates, economies, IS/LM framework, IS curves, LM curves, expansionary monetary policy, exogenous increase in domestic price levels, exogenous increase in savings, retirement, government spending, personal income tax, deficits.

Q24 Company assets, depreciation basis, marginal tax rates, after-tax rates, present value of tax shields.

Q25 Competitive firm, short-run operation, total revenue, total costs, fixed costs, average costs, marginal costs, average variable costs.

Q26 Investment, annual effective interest rates, accumulated values.

Q27 Stock prices, marginal requirements, marginal debts, interest, annual effective rates, dividends, return, short sale.

Q28 Monopolies, demand, marginal costs, prices, quantities, demand curves, continuous quantities.

Q29 Monopolies, marginal propensity to consume, income tax rates, government expenditure multiplier, exponential functions.

Q30 Stock prices, one-period put, exercise prices, risk-free rates, unexercised prices.

Q31 Investment, time-weighted returns, dollar-weighted returns.

Q32 Short-run supply curves, competitive industries, prices, industry output, production increase, industry supply curve, elastic supply curve, marginal cost curve, factor-price effect, shift down of the marginal cost curve.

Q33 Current liabilities, long-term liabilities, shareholder equity, total assets, EBIT [earnings before income and taxes], depreciation, interest, taxes, payout ratio, retained earnings, net income, dividends, internal growth rates.

Q34 Economies, goods, competitive supply and demand functions, prices, quantities, price ceilings, supply curve, deadweight loss, competitive equilibrium, consumer surplus, producer surplus, total surplus, excess demand.

Q35 Variance, equity returns, equal-weighted portfolio, beta, returns on assets, return on a market portfolio, slope, capital asset pricing model, derivatives (calculus).

Q36 Nominal exchange rates, inflation rates, real exchange rates.

Q37 Effective rates of interest, principals.

Q38   All-equity financed insurers, book value, return on equity, cash flow, annual earnings, dividends, free cash flow, opportunity costs, capital, discounted cash flow, plowback.

Q39   Debt ratio, debt beta, equity beta, expected return, risk-free interest rates, return on investment, target capital structure, risk, Modigliani-Miller capital structure theory, asset beta, capital asset pricing models.

Q40   Call option, common stocks, shares, standard deviations, continuous interest, compound interest, maturity of a call, risk-free rates, Black-Scholes, present value.

Q41   Bonds, semi-annual coupons, nominal yields, compound interest, annual effective interest rates, coupon payments, redemption value of bonds, annual effective yields, investment.

Q42   Supply and demand functions, prices, quantities, price elasticity of demand, initial equilibrium, percentage change, derivatives (calculus).

Q43   Market value, liabilities, debts, equity, beta, expected return, weighted average cost of capital, risk-free rates, expected risk premiums.

Q44   Earnings before interest and taxes, debt, corporate tax rate, dividend, average equity, return on average equity.

Q45   Force of interest, nominal rates of discount, convertible rates, accumulated value of funds, exponential functions, integrals.

Q46   Utility-maximizing consumers, indifference curves, utilities, slopes, budget curves.

Q47   Macro-economies, long-run view, real output, growth of real output, growth of inputs, velocity of money, growth in wage rates, wage-price spiral, inflation.

Q48   Stock prices, dividends, long-run dividend growth rates, capitalization rates, expected rates of return.

Q49   Investment, interest, nominal interest rates, convertible interest rates, simple interest, forces of interest, logarithmic functions.

Q50   Present value, annuities, perpetuity-immediate annuities, effective interest rates, annuity-immediate.

If we examine the frequency of some of the topics and techniques tested in this examination, we come up with the following result: price (16/50), marginal (13/50), cost (10/50), interest (10/50), growth (9/50), present value (9/50), cash flow (8/50), curve (8/50), effective (8/50), investment (7/50), market (7/50), debt (6/50), return (6/50), stock price (6/50), demand (5/50), dividend (5/50), equity (5/50), expected value (5/50).

Questions and Answers

Here are some examples from the May 2001 examination that show how some of these ideas and techniques were tested. The cited questions involve a variety of ideas, ranging from supply and demand, the business cycle, money supply, marginal tax rates, to effective interest rates, stock prices and the valuation of companies. The questions also use two special actuarial symbols: the symbol $a_{\overline{n}|}$, which stands for the value of an annuity of one dollar per year for n years, payable at the end of each year, and the symbol $a_{\overline{n}|i}$, denotes the value of an annuity of one dollar per year for n years at i percent interest per year, payable at the end of each year.

Question 3 *Suppose the economy is entering the contractionary phase of a business cycle. Which of the following is an indicator of this downturn in economic activity? (1) A decrease in the unemployment rate. (2) An increase in the number of new building permits for private housing units. (3) An increase in stock prices. (4) An increase in delivery lags. (5) An increase in business inventories.*

Answer An increase in business inventories indicates that demand is not as high as businesses anticipated, resulting in inventory accumulation. The decrease in demand is a reflection of the downturn in economic activity. ▲

Question 4 *A 20-year loan of 20,000 may be repaid under the following two methods: (1) Amortization method with equal annual payments at an annual effective rate of 6.5%, (2) Sinking fund method in which the lender receives an annual effective rate of 8% and the sinking fund earns an annual effective rate of j. Both methods require a payment of X to be made at the end of each year for 20 years. Calculate j.*

Answer We note that

$$X = \frac{20000}{a_{\overline{20}|0.065}} = 1815.13$$

Therefore,

$$1815.13 = \frac{20000}{s_{\overline{20}|j}} + (0.08)(20,000)$$

$$s_{\overline{20}|j} = 92.97$$

$$j = 14.18\%$$

This answers the question. ▲

Question 6 *Suppose a constant-cost, competitive industry is in long-run equilibrium. Now suppose the government imposes an annual licensing fee as* *a requirement for firms to produce in the industry. As a result of this fee, what will happen to the quantity supplied in the market and the quantity supplied by an individual firm in the long run?*

The possible answers are

1. The quantity supplied in the market will increase, and the quantity supplied by an individual firm will increase.
2. The quantity supplied in the market will increase, and the quantity supplied by an individual firm will decrease.
3. The quantity supplied in the market will decrease, but the quantity supplied by an individual firm will not change because some firms go out of business.
4. The quantity supplied in the market will decrease, and the quantity supplied by an individual firm will decrease.
5. The quantity supplied in the market will decrease, and the quantity supplied by an individual firm will increase.

Answer The licensing fee works the same as an increase in fixed costs; it shifts the market supply upward, increasing price and decreasing quantity demanded. At the firm level, however, it increases average costs without changing marginal costs; therefore, the representative firm increases output. This apparent paradox is resolved by the fact that in the long run some firms will go out of business. ▲

Question 12 *Bruce and Robbie each open up new bank accounts at time 0. Bruce deposits 100 into his bank account, and Robbie deposits 50 into his. Each account earns an annual effective discount rate of d. The amount of interest earned in Bruce's account during the 11th year is equal to X. The amount of interest earned in Robbie's account during the 17th year is also equal to X. Calculate X.*

Answer Bruce's interest in the 11th year is

$$\frac{100}{(1-d)^{10}} \left[\frac{1}{(1-d)} - 1 \right] = X$$

and Robbie's interest in the 17th year is

$$\frac{50}{(1-d)^{16}} \left[\frac{1}{(1-d)} - 1 \right] = X$$

$$= \frac{100}{(1-d)^{10}} \left[\frac{1}{(1-d)} - 1 \right]$$

$$(1-d)^6 = \frac{1}{2} \implies d = 10.91\%$$

$$X = \frac{100}{(1-0.1091)^{10}} \left[\frac{1}{1-0.1091} - 1 \right] = 38.88$$

This answers the question. ▲

Question 16 *The money supply is determined by the combined actions of the central bank, the commercial banking system, and the public's preferences regarding how they hold money. Which of the following will result in an increase in the money supply? (1) An increase in the public's demand for currency. (2) An exogenous increase in market interest rates. (3) The central bank increases the discount rate. (4) The central bank increases the reserve requirement ratio. (5) The central bank sells bonds to the public.*

Answer An increase in market interest rates will result in banks lending out excess reserves, which lowers free reserves and increases the money supply. ▲

Question 17 *At an annual effective interest rate of i, i > 0%, the present value of a perpetuity paying 10 at the end of each 3-year period, with the first payment at the end of year 6, is 32. At the same annual effective rate of i, the present value of a perpetuity-immediate paying 1 at the end of each 4-month period is X. Calculate X.*

Answer We note that

$$v^3 \frac{10}{(1+i)^3 - 1} = 32$$

Therefore,

$$10v^3 = 32(1+i)^3 - 32$$

Multiplying both sides by $(1+i)^3$ yields

$$10v^3(1+i)^3 = 32(1+i)^6 - 32(1+i)^3$$

and since $v^3(1+i)^3 = 1$, we have

$$0 = 32(1+i)^6 - 32(1+i)^3 - 10$$

This tells us that

$$(1+i)^3 = \frac{32 \pm \sqrt{2304}}{64} = 1.25$$

$$i = 7.72\%$$

$$X = \frac{1}{(1+i)^{1/3} - 1}$$

$$= \frac{1}{(1.0772)^{1/3} - 1} = 39.84$$

This answers the question. ▲

Question 22 *A company invests* 20,000 *in a project. The project is expected to have cash flows of* 3000 *at the end of each year for* 15 *years, with the first cash flow expected one year after the initial investment. Using the project's after-tax weighted average cost of capital, the project has a net present value of* 2496.27. *The following gives additional information about the company: (1) The company is financed with 40% equity and 60% debt. (2) The company's marginal tax rate is 25%. (3)* $r_E = 2r_D$, *where* r_E *is the cost of equity and* r_D *is the cost of debt. Calculate* r_E.

Answer Let i denote the after-tax weighted average of capital. Then

$$3000 a_{\overline{15}|i} - 20,000 = 2496.27$$

Therefore, $a_{\overline{15}|i} = 7.49876$. Hence $i = 10.25\%$. It follows that

$$10.25 = r_E (0.4) + r_D (1 - T_x) (0.6)$$
$$= 0.4 \cdot r_E + \frac{1}{2} r_E (1 - 0.25) (0.6) = 0.4 r_E + 0.225 r_E.$$

Hence $r_E = 16.4\%$. ▲

Question 24 *A company has an asset with a depreciation basis of* 100,000 *which can be depreciated by the following schedule:*

Year	Percent
1	33.33
2	44.45
3	14.81
4	7.41

The marginal tax rate is 35% and the pretax borrowing rate is 12%. Calculate the present value of the tax shields created by the depreciation.

Answer From the given information we conclude that

	Year 1	Year 2	Year 3	Year 4
Dollar deductions	33,330	33,340	14,810	7,410
Tax shields	11,666	15,558	5,184	2,594

and the after tax rate is $0.12(0.65) = 0.078$. Hence

$$
\begin{aligned}
PV &= 11{,}666/1.078 + 15{,}558/1.078^2 + 5{,}184/1.078^3 + 2{,}594/1.078^4 \\
&= 30{,}267
\end{aligned}
$$

This answers the question. ▲

Question 30 *A stock price can go up by 20% or down by 15% over the next period. The current stock price is greater than 70. You own a one-period put on the stock. The put has an exercise price of 78.26. The risk-free rate is 11.25%. If the put is exercised today, the amount received will be X. The price of the put today (unexercised) is also X. Calculate the current stock price.*

Answer Let p be the probability that a stock price goes up. Then

$$20p + (-15)(1 - p) = 11.25 \rightarrow p = 0.75$$

Moreover, the put value is
$$78.26 - X$$
if exercised now, and
$$\frac{0.75 \cdot 0 + 0.25 \cdot (78.26 - 0.85X)}{1.1125}$$

if not exercised now. By equating these two expressions and solving for X, we get $X = 75$. ▲

Question 32 *Which of the following statements about the short-run supply curve for a competitive industry is false? (1) As price rises, industry output goes up because firms in the industry increase production. (2) As price rises, firms not previously producing will start up production and thereby further increase industry output. (3) As price rises, entry of new firms tends to make the industry supply curve more elastic than the supply curve of typical firms in the industry. (4) As price and output increase for the industry, the factor-price effect is likely to make the industry supply curve less elastic. (5) As price and output increase for the industry, the marginal cost curve of each firm in the industry will likely shift down because of the factor-price effect.*

Answer As a result of the factor-price effect, the marginal cost curves of the firms do not shift down but up. ▲

Question 33 *You are given:*

Current Liabilities	300	*EBIT*	400
Long-term Liabilities	700	*Depreciation*	100
Shareholder Equity	1400	*Interest*	50
Total Assets	2400	*Taxes*	60

The company's payout ratio is 10%. *Determine the company's internal growth rate.*

Answer The following calculations show that the internal growth rate is 10.875% :

1. Assets = Liabilities + Shareholder Equity:

$$300 + 700 + 1400 = 2400$$

2. Net income = EBIT − Interest − Taxes:

$$400 - 50 - 60 = 290$$

   where the depreciation has already been subtracted to get EBIT.

3. Payout Ratio:

$$0.10 \ = \ \frac{\text{Dividends}}{\text{Net Income}}$$
$$= \ \frac{\text{Dividends}}{290}$$

   Therefore,
$$\text{Dividends} = 29$$

Putting it all together, we have

$$\text{Retained Earnings} \ = \ \text{Net Income} - \text{Dividends}$$
$$= \ 290 - 29 - 261$$

and

$$\text{Internal Growth Rate} \ = \ \frac{\text{Retained Earnings}}{\text{Assets}}$$
$$= \ \frac{261}{2400} = 10.875\%$$

This answers the question. ▲

Question 38 *You are the chief actuary for a small, all-equity financed insurer. The current book value of equity is 1000. In years 1 and 2, you will earn a return on equity (ROE) of 20% and reinvest all earnings. Starting in year 3 (and every year thereafter), your company's ROE will be 15%, your free cash flow will be 50% of annual earnings, and you will pay a dividend equal to 100% of free cash flow. You have been approached by another insurer who would like to buy your company. Assuming an opportunity cost of capital equal to 15%, use discounted cash flow to find the value of your company.*

	Y1	Y2	Y3	Y4	Y5
Book Equity	1000	1200	1440	1548	1664.1
ROE	20%	20%	15%	15%	15%
Earnings	200	240	216	232.2	249.62
Dividends	0	0	108	116.1	124.81
Plowback	200	240	108	116.1	124.81
Free Cash Flow	0	0	108	116.1	124.81

Answer

Starting in Year 3, we have

$$\text{Dividend Growth Rate} = \text{Plowback} \cdot \text{ROE}$$
$$= (0.5)(0.15) = 0.075$$

Therefore,

$$PV @ t = 2 \text{ of Future Dividends} = PV @ t = 2 \text{ of Free Cash Flow}$$
$$= 108/(0.15 - 0.075) = 1440$$

and

$$PV @ t = 0 \text{ of Free Cash Flow}$$
$$= 1440(1.15)^{-2} \approx 1089$$

This answers the question. ▲

Question 40 *You are interested in purchasing a call option on a common stock that is currently trading at a price of 100 per share. You are given the following information: (1) The standard deviation of the continuously compounded annual rate of return on the stock is 0.4. (2) The time to maturity of the call is three months. (3) At the risk-free rate,*

$$\ln\left(\frac{Current\ Share\ Price}{Present\ Value\ of\ the\ Exercise\ Price}\right) = -0.08.$$

Calculate the price of each call option using Black-Scholes.

Answer The Black-Scholes model tells us that the price of a call option is the difference between the expected benefit from acquiring the stock outright and the present value of paying the exercise price on the expiration days. In other words,

$$Price = CP \times N(d_1) - PV \times N(d_2),$$

where the current price CP is 100, and where the present value PV of the exercise price is 108.33, since at the risk-free rate and a current share price of 100, the log of the ratio of the current share price and the present value of the exercise price is 0.08. Moreover, $d_1 = -0.3$ and $d_2 = 0.5$, since $t = 0.25$ and $\sigma = 0.4$, so that $\sigma \times \sqrt{(t)} = 0.2$. Therefore,

$$
\begin{aligned}
Price &= 100 \times N(-0.3) - 108.33 \times N(-0.5) \\
&= 100 \times 0.3821 - 108.33 \times 0.3085 \\
&= 4.79
\end{aligned}
$$

This answers the question. ▲

Question 42 *The supply and demand functions for a good are $P = 1 + 4Q$ and $P = 4 - 2Q$, respectively, where P is price and Q is quantity. Now suppose an increase in the price of an input causes the supply function to become*

$$P = 2 + 4Q$$

What is the price elasticity of demand at the initial equilibrium?

Answer The correct answer follows from the definition of the price elasticity of demand. The percentage change in price from the initial equilibrium is $1/9$, and the percentage change in quantity demanded is $-1/3$; hence the price elasticity of demand is -3.00. ▲

Question 48 *A company's stock is currently selling for 28.50. Its next dividend, payable one year from now, is expected to be 0.50 per share. Analysts forecast a long-run dividend growth rate of 7.5% for the company. Tomorrow the long-run dividend growth rate estimate changes to 7%. Calculate the new stock price.*

Answer Current capitalization rate is

$$P_0 = DIV_1/(r - g)$$

In other words,
$$28.50 = 0.50/(r - 0.075)$$

Therefore, $r - 0.075 = 0.50/28.50 = 0.0175$, so that $r = 0.0925$. When the long-run growth rate changes, current price should adjust to reflect this change, and to keep the expected rate of return constant. This tells us that

$$P_0 = 0.50/(0.0925 - 0.07) = 22.22$$

This answers the question. ▲

2.6   SOA and CAS Course 3

This course deals with the use of actuarial models. In the Basic Education Catalog, Spring 2003 (*see* [4]), the Society of Actuaries describes the learning objective of the course by saying that "this course develops the candidate's knowledge of the theoretical basis of actuarial models and the application of those models to insurance and other financial risks." The word "model" is used by scientists for the tools they have developed to describe and explore their environments. Physicists build models to understand the universe around us, biologists build models to understand the long-term dynamics of interacting populations, economists build models to understand the interaction of the supply and demand of consumer goods, and actuaries build models to analyze the profitability of insurance plans, pension plans, and the returns on investment portfolios. Most scientists use their models to predict some aspects of the future and to provide a basis on which decisions can be made. These decisions can be ecological, economic, commercial and financial. In the case of actuaries, the decisions are usually financial.

In collaboration with Wolfram Research, the makers of the Mathematica software package, and ACTEX, the providers of actuarial study tools, Bruce Jones of the University of Western Ontario has developed a beautiful interactive course for studying actuarial models and building them (*see* Reference 10, Appendix F). This course is ideal for preparing the examinations in Courses 3 and 4. Jones begins his exposition by pointing out that "from the perspective of the actuary, a model can be defined as *a mathematical representation of a phenomenon*. This phenomenon usually has financial implications. Examples of phenomena that actuaries frequently model include the following: the time until death of an individual insured under a life insurance policy, the amount of insured losses under a health, automobile or property insurance policy, and the return of an investment portfolio."

From a mathematical point of view, a model can be many things. Among the familiar models are graphs that help us visualize quantitative relationships and functions that capture changes in a phenomenon and allow us to make predictions about the future.

   The starting points for building actuarial models are historical data and probabilistic assumptions. For example, a frequently encountered mathematical model
in actuarial science is the Poisson probability distribution. It is used, for example,
to model phenomena such as the number of automobile accidents at a particular
intersection in a city over a fixed period of time.

   The basic idea from statistics needed here is that of a *random variable*. What
are random variables? In an attempt to simplify their definition, many authors
have different ways of defining them. One author writes that "a random variable
is a variable whose values are determined by chance." Another writes that "a
random variable is a real-valued function for which the domain is a sample space."
At some point, all authors distinguish between discrete and continuous random
variables. The common element here is that a random variable is above all a *real-
valued function*. Moreover, its domain has a certain structure that statisticians
refer to as a *sample space*. The elements of the sample space are known as *sample
points* and sets of sample points are called *events*. In Ott and Longnecker's book
on statistical methods (*see* Reference 17, Appendix F), the random variables of
interest in Course 3 are called *quantitative random variables*. In the discrete
case, they allow us to introduce quantitative measures such as means and standard
deviations over their range of values. The key actuarial idea associated with a
random variable is that of the *expected value* of the variable.

Example 1 *Discrete random variable*

If you take two coins and list the number of possible head-tail combinations which
can be obtained by tossing the coins, the result is a sample space S. The events
are

$$e_1 = HH, e_2 = HT, e_3 = TH, e_4 = TT$$

The function $Y : S \to \mathbb{R}$ defined by

$$Y(e_1) = 0, Y(e_2) = Y(e_3) = 1, Y(e_4) = 2$$

is a random variable. It counts the number of heads of each sample points. Since
the domain of Y is finite, the function is called *discrete*. ▲

   The next required idea is that of a *probability distribution* of a random variable. It assigns to each value x of the random variable X a probability $0 \leq
p(x) \leq 1$, also denoted by $P(X = x)$, that measures the likelihood that the value
x is attained. It is assumed that the sum all values $p(x)$ over the domain of X is 1.

Example 2 *Continuous random variable*

The change in earnings per share of a particular stock over a fixed period of time
is a random variable. The sample space is an interval on the real line marking off

the time period over which the change is measured. The variable is continuous since it can take on arbitrary real numbers as values at all points of time in the interval. ▲

Example 3 *Expected value*

Suppose that you would like to insure your laptop computer for $2,000 against theft for one year. Suppose further that an insurance company has empirical evidence that the probability of having the laptop stolen in the first year is 1/10. What is your expected return from the insurance company if the premium you are charged is $100?

You have a chance of 1/10 of receiving $1,900 from the insurance company since you have already paid the company $100 in premiums. On the other hand, you have a chance of 9/10 of losing the $100 you have paid. The expected value of the probability distribution for X is

$$E\left(X\right) = (1900) \times (1/10) + (-100) \times (9/10) = 100$$

This means that if you insure your computer with the given company over a number of years, you will have an average net gain of $100 per year. The expected value from the insurance company's point of view, on the other hand, is

$$E\left(Y\right) = (-1900) \times (1/10) + (100) \times (9/10) = -100$$

In other words, the company can expect to lose $100 on average on this policy. ▲

Probability distributions are important statistical tools for analyzing the properties of random variables. In actuarial science, the binomial distributions and their associated Poisson and normal distributions play an important role.

Definition 4 *The function $P\left(y\right)$ is actually the limiting value of the widely known binomial probability distribution. It is derived from the binomial distribution by noting that*

$$\lim_{n \to \infty} \left(1 - \frac{\lambda}{n}\right)^n = e^{-\lambda}$$

and that therefore, with $\lambda = np$,

$$\lim_{n \to \infty} \binom{n}{y} p^y (1-p)^{n-y} = \frac{\lambda^y}{y!} e^{-\lambda}$$

Let us illustrate the descriptive and predictive aspect of a mathematical model by recalling a classical illustration of the binomial distribution due to Weldon. For historical details, we refer to the beautiful article on probability in the 1911 edition of the *Encylopedia Britannica* (*see* Reference 9, Page 394).

Example 5 *The Weldon experiment*

Suppose that we have n independent events, that the probability of a successful outcome of an event is p, and the probability of an unsuccessful event is q. If N represents the number of trials, then the formula

$$N(p+q)^n$$

counts the probable frequencies of the different results in a given number of trials.

Suppose that twelve dice are thrown a certain number of times, and that each face showing a 4, 5, or 6 is considered a success, whereas each face showing a 1, 2, or 3 is considered a failure. Then the probabilities of success and failure of each throw of the twelve dice is 1/2. Then

$$N\left(\frac{1}{2}+\frac{1}{2}\right)^{12}$$

where N is the total number of throws. Moreover, the binomial expansion of the expression

$$\left(\frac{1}{2}+\frac{1}{2}\right)^{12}$$

yields $A+B$, where

$$A = \frac{1}{4,096} + \frac{12}{4,096} + \frac{66}{4,096} + \frac{220}{4,096} + \frac{495}{4,096} + \frac{792}{4,096}$$

and

$$B = \frac{924}{4,096} + \frac{792}{4,096} + \frac{495}{4,096} + \frac{220}{4,096} + \frac{66}{4,096} + \frac{12}{4,096} + \frac{1}{4,096}$$

Let $N = 4,096$. Then

$$4,096\,(A+B)$$

is the sum of the theoretical frequencies of the different possible successes of 4,096 throws of twelve dice. The following table compares these frequencies with the experimental frequencies found by Weldon:

Successes	Observed Frequencies	Theoretical Frequencies
0	0	1
1	7	12
2	60	66
3	198	220
4	430	495

Successes	Observed Frequencies	Theoretical Frequencies
5	731	792
6	948	924
7	847	792
8	536	495
9	257	220
10	71	66
11	11	12
12	0	1
Total	4,096	4,096

We can see that the relationship between the two distributions is very close. From an actuarial point of view, one of the important properties of the binomial distribution is the fact that it is a building block for the Poisson distribution.

Here are a number of typical examples illustrating the use of the binomial, the Poisson, and the normal distribution:

Example 6 *A binomial experiment*

Suppose that a student is writing a multiple-choice examination consisting of 40 questions, each with five possible choices. Calculate the probability that the student guesses exactly 20 right answers.

Answer The probability of success in a binomial experiment with x successes in n trials is given by the formula

$$P(x) = \frac{n!}{(n-x)!x!}p^x q^{(n-x)}$$

where p is the probability of success in a single trial, and q is the probability of failure in a single trial. Since $n = 40$, $x = 20$, and $p = \frac{1}{5}$, we have

$$P(20) = \frac{40!}{20!20!}\left(\frac{1}{5}\right)^{20}\left(\frac{4}{5}\right)^{20} \approx 1.6665 \times 10^{-5}$$

This answers the question. ▲

Example 7 *A Poisson experiment*

In a clinical trial, 1,000 patients were treated with a new drug. Suppose that the known probability p of a person experiences negative side effects is 0.0025.

What is the probability that none of the 1,000 patients participating in the trial experience negative side effects?

Answer According to the Poisson formula, the probability of y successes in n trials is given by the formula

$$P(y) = \frac{\lambda^y}{y!} e^{-\lambda}$$

where $y = 0$ and $\lambda = np = 1000 \times 0.0025 = 2.5$. Therefore,

$$P(0) = \frac{2.5^0}{0!} e^{-2.5} \approx 0.082085$$

This answers the question. ▲

   This example could of course have been answered more directly by noting that the question is asking for the probability of $1,000$ failures in $1,000$ trials of an experiment with success rate 0.0025. This is just

$$(1 - .0025)^{1000} \approx 0.81828$$

Example 8 *A normal approximation*

Whatever its beauty and theoretical correctness as a model for statistical analysis, the binomial distribution is often computationally too complex for practical use. Consider the following problem, discussed in detail in Ott and Longnecker's book on statistical methods (*see* Reference 17, Appendix F, Page 182). A thousand voters are polled to determine their opinion on a municipal merger. What is the probability that 460 or fewer of them favor the merger if it is assumed that 50% of the entire population favors the change?

Answer In the binomial experiment, $n = 1,000$, and the probability $p = 1/2$. To answer the questions, we must compute the sum

$$P = P(460) + P(459) + \cdots + P(1) + P(0)$$

where
$$P(460) = \frac{1000!}{460!540!} \left(\frac{1}{2}\right)^{460} \left(\frac{1}{2}\right)^{540}$$

and so on. The number of calculations required for the solution is enormous. What is our way out? The central limit theorem for sums (*see* Reference 17, Appendix F, for example) enables us to approximate P using an approximate normal curve as an approximation to the required binomial distribution.

   The graphs of the functions

$$f(y) = \frac{1}{\sqrt{2\pi}\sigma} e^{-\frac{(y-\mu)^2}{2\sigma^2}}$$

are bell-shaped curves known as *normal curves*, whose shape depends on the parameters μ and σ. Moreover, the area

$$P(a \leq Y \leq b) = \int_a^b f(y)\, dy$$

under the curve of f can be interpreted as a probability.

If μ is the mean and σ the standard deviation of a normally distributed random variable Y with density function $f(y)$, then the probability that a randomly chosen value of Y will lie between a and b is $P(a \leq Y \leq b)$.

It is explained in Ott and Longnecker's book on statistical methods (*see* Reference 17, Appendix F, Page 184) that for large n and p not near 0 or 1, the distribution of a binomial random variable y may be approximated by a normal distribution with $\mu = np$ and $\sigma^2 = np(1-p)$, provided that $np \geq 5$ and $n(1-p) \geq 5$. The polling problem can be solved using the normal distribution since $np = 1000 \times .5 = 500 = n(1-p) \geq 5$.

Since most integrals involved in normal distribution problems have no closed-form solutions, approximate values of the integrals have been tabulated. For this purpose, an additional simplification has been introduced. Every normal distribution can be converted to standard form by letting

$$z = \frac{y - \mu}{\sigma}$$

and looking the value of z up in a table for standard normal curve areas.

If $\mu = np = 500$ and $\sigma = \sqrt{np(1-p)} = \sqrt{250} = 15.811$, then

$$z = \frac{y - \mu}{\sigma} = \frac{460 - 500}{15.811} \approx -2.53$$

Table 1 in Appendix of Ott and Longnecker's book on statistical methods (*see* Reference 17, Appendix F) tells us that the area under the normal curve to the left of 460 (for $z = -2.53$) is 0.0057. Therefore, the probability of observing 460 or fewer favoring the merger is about 0.0057. ▲

Ideas and Techniques

According to the SOA, "this course develops the candidate's knowledge of the theoretical basis of actuarial models and the application of those models to insurance and other financial risks. A thorough knowledge of calculus, probability and interest theory is assumed. A knowledge of risk management at the level of Course 1 is also assumed. The candidate will be required to understand, in an actuarial context, what is meant by the word 'model,' how and why models are used, their advantages and their limitations. The candidate will be expected to understand what important results can be obtained from these models for the purpose of making business decisions, and what approaches can be used to determine these results."

Examination Topics

The 2001 examination consisted of forty multiple-choice questions. It dealt with the following topics from the SOA and CAS syllabus:

Q1   Survival function, de Moivre's law, limiting age, integrals.

Q2   Term insurance, benefits, premiums, loss random variable.

Q3   Random variables, gamma distributions, variances, means, Poisson distributions, expected values, negative binomial distributions.

Q4   Poisson distributions, random variables, variances, compound distributions, independent processes.

Q5   Life insurance, whole life policy, level annual benefit premiums, benefit reserves.

Q6   Multiple decrement models, life tables, exponential functions, probabilities, integrals.

Q7   Probability models, probabilities, expected value, Markov chain.

Q8   Stock prices, geometric Brownian motion, drift coefficients, mean, variance, inverse transform method, uniform distribution, random numbers, exponential functions.

Q9   Life insurance, fully discrete insurance, annual benefit premiums, life expectancy.

Q10 Multiple decremental models, expected values, exponential functions, logarithmic functions, integrals.

Q11 Term insurance, death benefits, inverse transform method, present value random variable, uniform distributions.

Q12 Life insurance, death and surrender benefits, mortality tables, surrender rates, inverse transform method, policy terminated by death, policy terminated by surrender, uniform distributions, random variable indicating death, random variable indicating lapse of policy.

Q13 Time-until-death, hyperbolic assumption at fractional ages, independent lives, probabilities.

Q14 Life-table functions, force of mortality, mortality graphs.

Q15 Automobile insurance, negative binomial distributions, means, variances, Poisson distributions, gamma distributed means, variance of gamma distributions.

Q16 Probability distributions, mean, variance, independence, normal approximations, expected values.

Q17 Term insurance, present value random variable, death benefits, actuarial present value.

Q18 Endowment insurance, discrete insurance, death benefits, maturity benefits, level annual benefit premiums, benefit reserves, actuarial present value, future benefits.

Q19 Stop-loss insurance, independence, loss distribution, deductibles, actuarial expected value.

Q20 Insurance claims, compound Poisson claims process, probability, moment-generating functions, continuous premium rates, adjustment coefficients, exponential functions, expected values.

Q21 Markov process, insurance claims, probabilities of claims, independence, dividends, probability of failure.

Q22 Markov process, insurance claims, probabilities of claims, independence, expected dividends.

Q23 Continuous two-life annuities, continuous single life annuities, actuarial present value.

Q24 Disability insurance, length of payment random variable, gamma distributions, actuarial present value, improper integrals, exponential functions.

Q25 Discrete probability distributions, recursion relations, Poisson distributions, exponential functions, factorial function.

Q26 P/C insurance, loss models, aggregate loss, compound Poisson distributions, expected value, exponential distributions, deductibles, memoryless property.

Q27 Mortality models, expected number of survivors, uniform distribution of deaths (UDD), constant force assumptions.

Q28 Time-until-death, force of mortality, uniform distributions, probability of death, expected value, improper integrals, exponential functions.

Q29 Loss models, probability distributions, expected value, standard deviation, variance.

Q30 Stop-loss insurance, security loads, probability distributions, deductibles, expected value, sums of independent random variables.

Q31 Term insurance, level benefit premiums, benefit reserves, actuarial present value.

Q32 Whole life insurance, fully continuous insurance, level premiums, equivalence principle, death benefits, interest rates, loss random variable, future lifetime random variable.

Q33   Mortality models, uniformly distributions, complete-expectation-of-life, integrals.

Q34   Whole life insurance, death benefits, premiums, mortality, life tables, minimum annual rates of return, investments.

Q35   Whole life insurance, actuarial present value, force of mortality, death benefits, future lifetimes, independence, common stock model.

Q36   Poisson distributions, mean, probability, variance, compound distributions, expected value.

Q37   Poisson process, intensity functions, independence, distributed random variables, uniformly distributed claims, number of claims as a random variable, conditional expected value, integrals.

Q38   Whole life insurance, probabilities, death benefits, level premiums, independence, mortality, life tables, equivalence principle, benefit reserves, actuarial present value, future benefits.

Q39   Annuities, mortality, life tables, independence, normal approximations, present value random variables, lives, standard deviations.

Q40   Whole life insurance, death benefits, benefit premiums, mortality, life tables.

If we examine the frequency of some of the topics and techniques tested in this examination, we come up with the following result: insurance (23/40), benefit (19/40), random variable (19/40), distribution (18/40), expected value (14/40), function (12/40), probability (12/40), independence (10/40), mortality (9/40), premium (9/40), life (8/40), mean and standard deviation (8/40), exponential function (7/40), integration (7/40), variance (7/40), actuarial present value (6/40). The mathematical and statistical ideas in the course include binomial distributions, Poisson distributions, normal distributions, the analysis of stock prices, the calculation of insurance premiums, and ideas and situations. A detailed course description such as the one published by the Society of Actuaries its Basic Education Catalog, Fall 2002 (*see* Reference 3, Appendix F), for example, completes the picture.

Questions and Answers

Here are some examples from the May 2001 examination showing how some of these ideas and techniques were tested when the SOA and CAS examinations were still joint.

Question 1 *For a given life age* 30, *it is estimated that an impact of a medical breakthrough will be an increase of* 4 *years in*

$$\overset{\circ}{e}_{30}$$

the complete expectation of life. Prior to the medical breakthrough, s(x) followed de Moivre's law with w = 100 as the limiting age. Assuming de Moivre's law still applies after the medical breakthrough, calculate the new limiting age.

Answer By the de Moivre's law,

$$\overset{\circ}{e}_{30} = \int_0^{w-30} \left(1 - \frac{t}{w-30}\right) dt = \left[t - \frac{t^2}{2(\omega-30)}\right]_0^{\omega-30} = \frac{w-30}{2}$$

Prior to medical breakthrough, with $w = 100$, we therefore have

$$\overset{\circ}{e}_{30} = \frac{100-30}{2} = 35$$

After medical breakthrough,

$$\left(\overset{\circ}{e}\right)'_{30} = \overset{\circ}{e}_{30} + 4 = 39 = \frac{w'-30}{2}$$

It follows that $w' = 108$. ▲

Question 2 *On January 1, 2002, Pat, age 40, purchases a 5-payment, 10-year term insurance of 100,000: (1) Death benefits are payable at the moment of death. (2) Contract premiums of 4000 are payable annually at the beginning of each year for 5 years. (3) i = 0.05. (4) L is the loss random variable at time of issue. Calculate the value of L if Pat dies on June 30, 2004.*

Answer It follows from the given information that

$$_0L = 100,000v^{2.5} - 4000\ddot{a}_{\overline{3}|.05} = 77,079$$

This answers the question. ▲

Question 5 *For a fully discrete 20-payment whole life insurance of 1000 on (x), you are given: (1) i = 0.06. (2) $q_{x+19} = 0.01254$. (3) The level annual benefit premium is 13.72. (4) The benefit reserve at the end of year 19 is 342.03. Calculate 1000 P_{x+20}, the level annual benefit premium for a fully discrete whole life insurance of 1000 on (x + 20).*

Answer The given information tells us that

$$1000 \, {}_{20}^{20}V_x = 1000 A_{x+20}$$
$$= \frac{1000 \left({}_{19}^{20}V_x + {}_{20} P_x \right) (1.06) - q_{x+19} (1000)}{p_{x+19}}$$
$$= \frac{(342.03 + 13.72)(1.06) - 0.01254(1000)}{0.98746}$$
$$= 369.18$$

and, therefore,

$$\ddot{a}_{x+20} = \frac{1 - 0.36918}{0.06/1.06} = 11.1445$$

It follows that

$$1000 P_{x+20} = 1000 \frac{A_{x+20}}{\ddot{a}_{x+20}} = \frac{369.18}{11.1445} = 33.1$$

This answers the question. ▲

Question 7 *A coach can give two types of training, "light" or "heavy," to his sports team before a game. If the team wins the prior game, the next training is equally likely to be light or heavy. But, if the team loses the prior game, the next training is always heavy. The probability that the team will win the game is 0.4 after light training and 0.8 after heavy training. Calculate the long run proportion of time that the coach will give heavy training to the team.*

Answer Let "light training" be State 1 and "heavy training" be State 2. Then the probabilities P_{ij} involved are

$$P_{11} = 0.4 \times 0.5 + 0.6 \times 0 = 0.2$$
$$P_{12} = 0.4 \times 0.5 + 0.6 \times 1 = 0.8$$
$$P_{21} = 0.8 \times 0.5 + 0.2 \times 0 = 0.4$$
$$P_{22} = 0.8 \times 0.5 + 0.2 \times 1 = 0.6$$

and the transition matrix of the given Markov process is therefore

$$P = \begin{bmatrix} 0.2 & 0.8 \\ 0.4 & 0.6 \end{bmatrix}$$

Let π_1 be the long-run probability that light training will be given, and π_2 that heavy training will take place. Then we can tell from the matrix P that

$$\pi_1 = 0.2\pi_1 + 0.4\pi_2$$
$$\pi_2 = 0.8\pi_1 + 0.6\pi_2$$

We also know that $\pi_1 + \pi_2 = 1$. Hence

$$1 - \pi_2 = 0.2\,(1 - \pi_2) + 0.4\pi_2$$
$$= 0.2 + 0.2\pi_2$$

Therefore, $1.2\pi_2 = 0.8$ and $\pi_2 = \frac{8}{12} = \frac{2}{3}$. ▲

Question 9 (x) *and* (y) *are two lives with identical expected mortality. You are given that* $P_x = P_y = 0.1$, *that* $P_{\overline{xy}} = 0.06$, *where* $P_{\overline{xy}}$ *is the annual benefit premium for a fully discrete insurance of 1 on* $(\overline{xy})$, *and that* $d = 0.06$. *Calculate the premium* P_{xy}, *the annual benefit premium for a fully discrete insurance of 1 on* (xy).

Answer We note that $P_s = 1/\ddot{a}_s - d$, where s can stand for any of the statuses under consideration.

Therefore,

$$\ddot{a}_s = \frac{1}{P_s + d}$$

$$\ddot{a}_x = \ddot{a}_y = \frac{1}{0.1 + 0.06} = 6.25$$

$$\ddot{a}_{\overline{xy}} = \frac{1}{0.06 + 0.06} = 8.333$$

and since $\ddot{a}_{\overline{xy}} + \ddot{a}_{xy} = \ddot{a}_x + \ddot{a}_y$,

$$\ddot{a}_{xy} = 6.25 + 6.25 - 8.333 = 4.167$$

$$P_{xy} = \frac{1}{4.167} - 0.06 = 0.18$$

This answers the question. ▲

Question 10 *For students entering a college, you are given the following from a multiple decrement model: (1) 1000 students enter the college at* $t = 0$. *(2) Students leave the college for failure (1) or all other reasons (2). (3)* $\mu^{(1)}\,(t) = \mu$, $0 \le t \le 4$ *and* $\mu^{(2)}\,(t) = 0.04$, $0 \le t < 4$. *(4) 48 students are expected to leave the college during their first year due to all causes. Calculate the expected number of students who will leave because of failure during their fourth year.*

Answer It follows from the given information that

$$d_0^{(\tau)} = 1000 \int_0^1 e^{-(\mu+0.04)t} (\mu+0.04)\, dt$$

$$= 1000 \left(1 - e^{-(\mu+0.04)}\right) = 48$$

Hence, $e^{-(\mu+0.04)} = 0.952$. It follows that

$$\mu + 0.04 = -\ln(0.952) = 0.049,$$

and, therefore, that $\mu = 0.009$. This tells us that

$$d_3^{(1)} = 1000 \int_3^4 e^{-0.049t} (0.009)\, dt$$

$$= 1000 \frac{0.009}{0.049} \left(e^{-(0.049)(3)} - e^{-(0.049)(4)}\right) = 7.6$$

This answers the question. ▲

Question 15 *An actuary for an automobile insurance company determines that the distribution of the annual number of claims for an insured chosen at random is modeled by the negative binomial distribution with mean 0.2 and variance 0.4. The number of claims for each individual insured has a Poisson distribution and the means of these Poisson distributions are gamma distributed over the population of those insured. Calculate the variance of this gamma distribution.*

Answer Using the conditional mean and variance formulas, we get

$$E[N] = E_\Lambda(N \mid \Lambda)$$
$$\mathrm{Var}[N] = \mathrm{Var}_\Lambda(E(N \mid \Lambda)) + E_\Lambda(\mathrm{Var}(N \mid \Lambda))$$

Since N, given lambda, is just a Poisson distribution, these equations simplify to

$$E[N] = E_\Lambda(\Lambda)$$
$$\mathrm{Var}[N] = \mathrm{Var}_\Lambda(\Lambda) + E_\Lambda(\Lambda)$$

Using the given values $E[N] = 0.2$ and $\mathrm{Var}[N] = 0.4$, we therefore get

$$0.4 = \mathrm{Var}_\Lambda(\Lambda) + 0.2$$

It follows that $\mathrm{Var}_\Lambda(\Lambda) = 0.2$. ▲

Question 16 *A dam is proposed for a river which is currently used for salmon breeding. You have modeled: (1) For each hour the dam is opened the number of* *salmon that will pass through and reach the breeding grounds has a distribution with mean* 100 *and variance* 900. *(2) The number of eggs released by each salmon has a distribution with mean of* 5 *and variance of* 5. *(3) The number of salmon going through the dam each hour it is open and the numbers of eggs released by the salmon are independent. Using the normal approximation for the aggregate number of eggs released, determine the least number of whole hours the dam should be left open so the probability that* 10,000 *eggs will be released is greater than* 95%.

Answer Let N denote the number of salmon, X the eggs from one salmon, and S the total eggs. Then $E(N) = 100t$ and $\text{Var}(N) = 900t$.

Therefore,

$$E(S) = E(N)E(X) = 500t$$
$$\text{Var}(S) = E(N)\text{Var}(X) + E^2(X)\text{Var}(N)$$
$$= 100t \times 5 + 25 \times 900t = 23,000t$$

Hence,

$$P(S > 10,000) = P\left(\frac{S - 500t}{\sqrt{23,000t}} > \frac{10,000 - 500t}{\sqrt{23,000t}}\right) = .95$$

Therefore,

$$10,000 - 500t = -1.645 \times \sqrt{23000}\sqrt{t} = -250\sqrt{t}$$
$$40 - 2t = -\sqrt{t}$$
$$2\left(\sqrt{t}\right)^2 - \sqrt{t} - 40 = 0$$
$$\sqrt{t} = \frac{1 \pm \sqrt{1 + 320}}{4} = 4.73$$
$$t = 22.4 \approx 23$$

This answers the question. ▲

Question 18 *For a special fully discrete* 20*-year endowment insurance on* (55):

(1) *Death benefits in year k are given by* $b_k = (21 - k), k = 1, 2, \ldots, 20.$
(2) *The maturity benefit is* 1.
(3) *Annual benefit premiums are level.*
(4) $_kV$ *denotes the benefit reserve at the end of year k*, $k = 1, 2, \ldots, 20.$
(5) $_{10}V = 5.0.$
(6) $_{19}V = 0.6.$
(7) $q_{65} = 0.10.$
(8) $i = 0.08.$

Calculate $_{11}V$.

Answer Let π denote the benefit premium. Since $_{19}V$ is the difference between the actuarial present value of the future benefits and the actuarial present value of the future premiums, we have

$$0.6 = \frac{1}{1.08} - \pi$$

Therefore, $\pi = 0.326$. It follows that

$$_{11}V = \frac{(_{10}V + \pi)(1.08) - (q_{65})(10)}{p_{65}}$$

$$= \frac{(5.0 + 0.326)(1.08) - (0.10)(10)}{1 - 0.10}$$

$$= 5.28$$

This answers the question. ▲

Question 19 *For a stop-loss insurance on a three-person group we are given the following information: (1) Loss amounts are independent. (2) The distribution of loss amount for each person is:*

Loss Amount	Probability
0	0.4
1	0.3
2	0.2
3	0.1

(3) The stop-loss insurance has a deductible of 1 for the group. Calculate the net stop-loss premium.

Answer Let X denote the losses on one life. Then

$$E[X] = (0.3)(1) + (0.2)(2) + (0.1)(3) = 1$$

Now let S denote the total losses. It follows that

$$E[S] = 3E[X] = 3$$
$$E[(S-1)_+] = E[S] - 1(1 - F_s(0))$$
$$= E[S] - (1)(1 - f_s(0))$$
$$= 3 - (1)(1 - 0.4^3)$$
$$= 3 - 0.936$$
$$= 2.064$$

This answers the question. ▲

Question 20 *An insurer's claims follow a compound Poisson claims process with two claims expected per period. Claim amounts can be only $1, 2,$ or 3 and these are equal in probability. Calculate the continuous premium rate that should be charged each period so that the adjustment coefficient will be 0.5.*

Answer We have

$$M_x(r) = E\left[e^{rx}\right] = \frac{e^r + e^{2r} + e^{3r}}{3}$$

$$M_x(0.5) = \frac{e^{0.5} + e + e^{1.5}}{3} = 2.95$$

$$p_1 = E[X] = \frac{1 + 2 + 3}{3} = 2$$

$$\lambda[M_x(r) - 1] = cr$$

Since $\lambda = 2$ and $r = 0.5$,

$$2[M_x(0.5) - 1] = 0.5c$$
$$2(2.95 - 1) = 0.5c$$
$$3.9 = 0.5c$$
$$c = 7.8 = \text{ premium rate per period}$$

This answers the question. ▲

Question 24 *For a disability insurance claim, the claimant will receive payments at the rate of $20,000$ per year, payable continuously as long as she remains disabled. The length of the payment period in years is a random variable with the gamma distribution with parameters $\alpha = 2$ and $\theta = 1$. Payments begin immediately and $\delta = 0.05$. Calculate the actuarial present value of the disability payments at the time of disability.*

Answer We have

$$\bar{a} = \int_0^\infty \bar{a}_{\overline{t}|} f(t)\, dt = \int_0^\infty \frac{1 - e^{-0.05t}}{0.05} \frac{1}{\Gamma(2)} t e^{-t}\, dt$$

$$= \frac{1}{0.05} \int_0^\infty \left(t e^{-t} - t e^{-1.05t} \right) dt$$

$$= \frac{1}{0.05} \left[-(t+1)e^{-t} + \left(\frac{t}{1.05} + \frac{1}{1.05^2} \right) e^{-1.05t} \right] \Big|_0^\infty$$

$$= \frac{1}{0.05} \left[1 - \left(\frac{1}{1.05} \right)^2 \right] = 1.85941$$

and $20,000 \times 1.85941 = 37,188$. ▲

Question 31 *For a special fully discrete 3-year term insurance on* (x)*, the level benefit premiums are paid at the beginning of each year,* $i = 0.06$*, and*

k	b_{k+1}	q_{x+k}
0	200,000	0.03
1	150,000	0.06
2	100,000	0.09

Calculate the initial benefit reserve for year 2.

Answer Let π denote the benefit premium and define

$$A = (0.03)(200,000)v$$
$$B = (0.97)(0.06)(150,000)v^2$$
$$C = (0.97)(0.94)(0.09)(100,000)v^3$$

Then the actuarial present value of benefits is

$$A + B + C = 5660.38 + 7769.67 + 6890.08$$
$$= 20,320.13$$

Therefore, the actuarial present is

$$\ddot{a}_{x:\overline{3}|}\pi = \left[1 + 0.97v + (0.97)(0.94)v^2\right]\pi$$
$$= 2.7266\pi,$$

so that

$$\pi = \frac{20,320.13}{2.7266} = 7452.55$$

The initial benefit reserve for Year 2 is therefore

$$_1V + \pi = \frac{(7452.55)(1.06) - (200,000)(0.03)}{1 - 0.03} + 7452.55$$
$$= 9411.01$$

This answers the question. ▲

Question 36 *The number of accidents follows a Poisson distribution with mean 12. Each accident generates 1, 2, or 3 claimants with probabilities* $\frac{1}{2}, \frac{1}{3}, \frac{1}{6}$*, respectively. Calculate the variance in the total number of claimants.*

Answer We treat the variances as three independent Poisson variables, corresponding to 1, 2, or 3 claimants.

$$\text{rate}_1 = \tfrac{1}{2} \times 12 = 6 \qquad \text{Var}_1 = 6$$
$$\text{rate}_2 = 4 \qquad\qquad \text{Var}_2 = 4 \times 2^2 = 16$$
$$\text{rate}_3 = 2 \qquad\qquad \text{Var}_3 = 18$$

Therefore Var $= 6 + 16 + 18 = 40$, since independent. ▲

Question 37 *For a claims process, you are given: (1) The number of claims* $\{N(t), t \geq 0\}$ *is a nonhomogeneous Poisson process with intensity function*

$$\lambda(t) = \begin{cases} 1 & \text{if } 0 \leq t < 1 \\ 2 & \text{if } 1 \leq t < 2 \\ 3 & \text{if } 2 \leq t \end{cases}$$

(2) Claims amounts Y_i *are independently and identically distributed random variables that are also independent of* $N(t)$. *(3) Each* Y_i *is uniformly distributed on* $[200, 800]$. *(4) The random variable P is the number of claims with claim amount less than 500 by time* $t = 3$. *(5) The random variable Q is the number of claims with claim amount greater than 500 by time* $t = 3$. *(6) R is the conditional expected value of P, given* $Q = 4$. *Calculate R.*

Answer Since

$$\int_0^3 \lambda(t)\, dt = 6$$

it follows that $N(3)$ is Poisson with $\lambda = 6$. Moreover, P is Poisson with mean 3 (with mean 3 since $Prob(Y_i < 500) = 0.5$). Since P and Q are independent, the mean of P is 3, no matter what the value of Q is. ▲

| **2.7** | **CAS Course 3** |

Since November 2003, the SOA and CAS versions of Course 3 are no longer identical. The Casualty Actuarial Society has decided on a different focus for this course. In addition to the general objective of expecting candidates to be able to apply actuarial models to business applications, as identified in the SOA curriculum, the Casualty Actuarial Society has placed new emphasis on a list of specific types of model that are covered in the course:

Survival and Contingent Payment Models

"Candidates should be able to work with discrete and continuous univariate probability distributions for failure time random variables. They will be expected to set up and solve equations in terms of life table functions, cumulative distribution functions, survival functions, probability density functions, and hazard functions (e.g., force of mortality), as appropriate. They should have similar facility with models of the joint distribution of two failure times (multiple lives) and the joint distribution of competing risks (multiple decrement). They should be able to formulate and apply stochastic and deterministic models for the present value of a set of future contingent cash flows under an assumed interest rate structure. Candidates also should be able to apply the equivalence principle, and other principles in the text, to associate a cost or pattern of (possibly contingent) costs with a set of future contingent cash flows."

Frequency and Severity Models

"Candidates should be able to define frequency (counting) and severity distributions, and be able to use the parameters and moments of these distributions. Candidates also should be able to work with the families of distributions generated by algebraic manipulation and mixing of the basic distributions presented."

Compound Distribution Models

"Candidates should be able to calculate the probabilities associated with a compound distribution when the compounding distribution is one of the frequency distributions presented in the syllabus, and the compounded distribution is discrete or a discretization of a continuous distribution. Candidates also should be able to adjust such probability calculations for the impact of policy modifications such as deductibles, policy limits, and coinsurance."

Stochastic Process Models

"Candidates should learn to solve problems using stochastic processes. They also should learn how to determine the probabilities and distributions associated with these processes. The following stochastic processes will be covered: Markov chain (discrete-time and continuous-time) processes (*see* Reference 20, Appendix F, for example), counting processes, Poisson process (including nonhomogeneous and compound Poisson processes), and Brownian motion (*see* Reference 22, Appendix F)."

Ruin Models

"Candidates should be able to analyze the probability of ruin using various models. Other topics covered in this section include the determination of the characteristics of the distribution of the amount of surplus (deficit) at the first time below the initial level and at the lowest level (maximal aggregate loss), and the impact of reinsurance."

Simulation of Models

"Candidates should be able to generate discrete and continuous random variables using basic simulation methods. They also should be able to construct algorithms to simulate outcomes using stochastic models."

2.8 | SOA and CAS Course 4

Whereas Course 3 deals with an understanding of actuarial models, the learning objectives of Course 4 concern the building of such models. According to the Society of Actuaries (*see* Reference 3, Appendix F), Course 4 "provides an introduction to modeling and covers important actuarial and statistical methods that are useful in modeling. A thorough knowledge of calculus, linear algebra, probability and mathematical statistics is assumed. The candidate will be required to understand the steps involved in the modeling process and how to carry out these steps in solving business problems. The candidate should be able to: (1) analyze data from an application in a business context; (2) determine a suitable model including parameter values; and (3) provide measures of confidence for decisions based upon the model. The candidate will be introduced to a variety of tools for the calibration and evaluation of the models in Course 3."

Ideas and Techniques

In [10], Jones discusses basic types of actuarial models and provides an electronic tool for constructing and studying them. Among the problems discussed are stochastic models, in which given phenomena are represented in probabilistic terms and deterministic ones, where given events are assumed to occur with certainty. He also hints at other types of models, built with relatively new mathematical techniques. This certainly demonstrates the dynamic nature of actuarial science. New problems require new solutions, all the time. As is shown in the book on loss models by Klugman et al. (*see* Reference 12, Appendix F), stochastic models include loss models, survival models, contingent payment models, credibility models, linear regression models, stochastic processes, and time-series models. According to the Society of Actuaries syllabus (*see* Reference 3, Appendix F), "the candidate is expected to apply statistical methods to sample data to quantify and evaluate the models presented in Course 3 and to use the models to solve problems set in a business context."

Examination Topics

The 2001 examination consisted of 40 multiple-choice questions. It dealt with the following topics from the SOA and CAS syllabus:

Q1 Invertible autoregressive moving average ARMA models, time series, auto-correlation function, MA(1) process, quadratic equations.

Q2 Poisson distributions, means, prior distributions, probability density functions, exponential functions, variances, posterior distributions, factorial function, gamma distributions.

Q3 Auto insurance, randomly selected policies, kurtosis, μ, σ.

Q4 Auto insurance, randomly selected policies, product-limit estimates, survival probabilities, censored data.

Q5 Multiple regression, F-statistic, significant variables, regression coefficients.

Q6 Full credibility standard, expected claims, square-root rule, partial credibility, Bühlmann credibility formula, exposure units.

Q7 Loss models, exponential distributions, maximum likelihood estimates, exponential functions, logarithmic functions, derivatives (calculus).

Q8 Mortalities, reference hazard rates, cumulative relative excess mortalities.

Q9 Dickey-Fuller unit root test, unrestricted regressions, restricted regressions, significance levels, random walk hypothesis, F-distributions, critical values.

Q10 Risk models, claim size distributions, probabilities, independence of claims, Bayesian premiums, variance, expected value.

Q11 Risk models, claim size distributions, probabilities, independence of claims, Bühlmann credibility premiums, expected present value, variance of hypothetical means.

Q12 Random observations, probability density functions, Kolmogorov-Smirnov statistics, integrals.

Q13 Method of least squares, standardized coefficients.

Q14 Mortalities, right-censored data, improper integrals, Aalen estimates, standard deviations, Nelson-Aalen estimators, cumulative hazard functions.

Q15 Mortalities, right-censored data, symmetric confidence intervals, mean survival times, integrals.

Q16 Loss models, Weibull distributions, maximum likelihood estimates, exponential functions, Weibull density functions, logarithmic functions, derivatives (calculus).

Q17 Autoregressive moving average ARMA(1,1) models, time series, variance.

Q18 Auto insurance, claim frequencies, Poisson distributions, mean, prior distributions, probability density functions, expected number of claims, exact posterior densities, posterior means, improper integrals.

Q19 Chi-square test, Poisson distributions, means, expected number of observations, exponential functions, factorial function.

Q20 Maximum likelihood estimates, Poisson distributions, negative binomial distributions, negative loglikelihoods, likelihood ratio test, null hypothesis.

Q21 Loss models, independence, loss ratios, weighted least squares estimators, derivatives, minima.

Q22 Mortalities, cumulative hazard functions, log-rank test, significance levels, chi-square statistics.

Q23 Risk models, means, variance, expected value, annual process variances, Bühlmann-Straub credibility factor, limits at infinity.

Q24 Claims models, multiple regression, average claim costs, lognormal error components, inflation, linear models, logarithmic functions, exponential functions.

Q25 Loss models, loss ratios, standard deviations, delta method, partial derivatives, variance.

Q26 Auto insurance, random variables, time lag, probabilities, survival functions, right-truncated data.

Q27 Sales models, seasonal adjustments.

Q28 Insurance claims models, expected value, prior probabilities, posterior probabilities.

Q29 Claims models, variance, two-tailed rank-sum hypothesis test, probability distributions, p-values.

Q30 Loss models, exponential distributions, maximum likelihood estimates, deductibles, policy limits, expected payments per loss, inflation, scale parameters, sample means, exponential functions.

Q31 Proportional hazards regressions, Cox models, covariate vectors, partial likelihoods, exponential functions.

Q32 Loss models, nonparametric empirical Bayes credibility premiums, preservation of total losses.

Q33 Annual premium income, loss ratio, two-variable linear regression models, slope coefficients, least-squares estimators, error terms, autoregressive AR(1) models, auto-correlation coefficients, standard errors, biased download estimators, Cochrane-Orcutt procedure, consistent estimators of the model slope.

Q34 Automobile insurance, random variable describing the time lag in settling a claim, maximum likelihood estimates, truncated observations, loglikelihoods, derivatives (calculus).

Q35 Functions defined by cases, hazard rates, censored claims, kernel-smoothed estimates, bandwidth, biweight kernels, log-transformed confidence intervals.

Q36 Autoregressive moving average ARMA(p,q) models, time series, autocorrelation function, simulated series (time series generated by the model), residual of the model, white-noise process, residual autocorrelations, normally distributed random variables, means, variances, Q-statistics, chi-square distributions, degrees of freedom, large displacements.

Q37 Claims models, compensation coverage, Poisson distributions, uniform distributions, posterior probability, posterior distributions, normalizing constants, integrals, posterior density, exponential functions.

Q38 Claims models, compensation coverage, Poisson distributions, uniform distributions, Bühlmann credibility estimates, expected number of claims, variance, Poisson parameter.

Q39 Claims models, independent distributions, exponential distributions, means, standard deviations, second raw moment, component means, quadratic equations.

Q40 Two-variable regression, standard error.

If we examine the frequency of some of the topics and techniques tested in this examination, we come up with the following result: function (26/40), distribution (23/40), mean and standard deviation (14/40), estimate (13/40), variance (12/40), model (11/40), expected value (9/40), probability (9/40), loss (8/40), exponential function (7/40), random variable (7/40), integration (6/40), differentiation (5/40), independence (5/40), insurance (5/40), mortality (5/40). The list questions and answers involve time series, specific formula and distributions, such as the Weibull distribution (*see* [21]), random walks, multiple regression, the chi square test (*see* [17]), the gamma distribution (*see* [21]), confidence intervals, least-squares estimates, and other ideas and techniques. Here are some sample questions that illustrate the "look and feel" of an examination in Course 4:

Questions and Answers

Question 1 *You are given the following information about an invertible ARMA time-series model:*

$$\begin{cases} \rho_1 = -0.4 \\ \rho_k = 0 \qquad k = 2, 3, 4, \ldots \end{cases}$$

Determine θ_1.

Answer Because the autocorrelation function is zero starting with lag 2, this must be an $MA(1)$ model. Then

$$-.4 = \rho_1 = \frac{-\theta_1}{1 + \theta_1^2}$$

so that

$$-.4 - .4\theta_1^2 = -\theta_1.$$

Therefore,

$$.4\theta_1^2 - \theta_1 + .4 = 0$$

This quadratic equation has two roots, 0.5 and 2. Because the coefficient's absolute value must be less than 1, only 0.5 is acceptable. ▲

Question 6 *You are given that the full credibility standard is* 100 *expected claims and that the square-root rule is used for partial credibility. You approximate the partial credibility formula with a Bühlmann credibility formula by selecting a Bühlmann k value that matches the partial credibility formula when* 25 *claims are expected. Determine the credibility factor for the Bühlmann credibility formula when* 100 *claims are expected.*

Answer The number of expected claims (e) is proportional to the number of exposure units (n). Let $e = cn$. Using Bühlmann credibility and partial credibility gives:

$$\sqrt{\frac{25}{100}} = \frac{1}{2} = \frac{25/c}{25/c + k} = \frac{25}{25 + ck}$$

Therefore $ck = 25$. When we have 100 expected claims,

$$Z = \frac{100/c}{100/c + k} = \frac{100}{100 + ck} = \frac{100}{100 + 25} = 0.80$$

This answers the question. ▲

Question 9 *A Dickey-Fuller unit root test was performed on* 100 *observations of each of three price series by estimating the unrestricted regression*

$$Y_t - Y_{t-1} = \alpha + \beta t + (\rho - 1) Y_{t-1}$$

and then the restricted regression

$$Y_t - Y_{t-1} = \alpha.$$

You are given that

Price Series	Unrestricted Error Sums	Restricted Error Sums
I	3233.8	3552.2
II	1131.8	1300.5
III	211.1	237.0

and that the critical value at the 0.01 significance level for the F-distribution calculated by Dickey and Fuller is 5.47. For which series do you reject at the 0.10 significance level the hypothesis of a random walk?

Answer We know that the F statistics if given by the formula

$$F = (N - k) \frac{(ESS_R - ESS_{UR})}{q(ESS_{UR})}$$

and $N = 100$, $k = 3$, $q = 2$. Therefore,

Series I: $F = 97 \dfrac{(3552.2 - 3233.8)}{2(3233.8)} = 4.78$ (Fail to reject)

Series II: $F = 97 \dfrac{(1300.5 - 1131.8)}{2(1131.8)} = 7.23$ (Reject)

Series III: $F = 97 \dfrac{(237.0 - 211.1)}{2(211.1)} = 5.95$ (Reject)

This answers the question. ▲

Question 15 *For a mortality study with right-censored data, you are given:*

t_i	d_i	Y_i	$\dfrac{d_i}{Y_i(Y_i - d_i)}$	$\widehat{S}(t_i)$	$\int_{t_i}^{\infty} \widehat{S}(t)\, dt$
1	15	100	0.0018	0.8500	14.424
8	20	65	0.0068	0.5885	8.474
17	13	40	0.0120	0.3972	3.178
25	31	31	—	0.0000	0.000

Determine the symmetric 95% confidence interval for the mean survival time.

Answer By the definition of $\widehat{\mu}_\tau$ we have

$$\widehat{\mu}_\tau = \int_0^\tau \widehat{S}(t)\,dt$$
$$= (1.0 \times 1) + (0.85 \times 7) + (0.5885 \times 9) + (0.3972 \times 8) = 15.42$$

Therefore,

$$\widehat{V}[\widehat{\mu}_\tau] = \sum_{i=1}^D \left[\int_{t_i}^\tau \widehat{S}(t)\,dt \right]^2 \frac{d_i}{Y_i(Y_i - d_i)}$$
$$= 14.424^2 \times 0.0018 + 8.474^2 \times 0.0068 + 3.178^2 \times 0.0120$$
$$= 0.9840$$

We conclude that the 95% confidence interval is $15.42 \pm 1.96 \times \sqrt{0.9840}$. ▲

Question 16 *A sample of ten losses has the following statistics:*

$$\sum_{i=1}^{10} X^{-2} = 0.00033674 \qquad \sum_{j=1}^{10} X^{0.5} = 488.97$$

$$\sum_{i=1}^{10} X^{-1} = 0.023999 \qquad \sum_{j=1}^{10} X = 31{,}939$$

$$\sum_{i=1}^{10} X^{-0.5} = 0.34445 \qquad \sum_{j=1}^{10} X^2 = 211{,}498{,}983$$

You assume that the losses come from a Weibull distribution with $\tau = 05$. Determine the maximum likelihood estimate of the Weibull parameter θ.

Answer The Weibull density function is $f(x) = .5(x\theta)^{-.5} e^{-(x/\theta)^{.5}}$. Therefore the likelihood function is

$$L(\theta) = \prod_{j=1}^{10} .5(x_j\theta)^{-.5} e^{-(x_j/\theta)^{.5}}$$
$$= (.5)^{10} \left(\prod_{j=1}^{10} x_j \right)^{-.5} \theta^{-5} e^{-\theta^{-.5}\sum_{j=1}^{10} x_j^{.5}}$$
$$\propto \theta^{-5} e^{-488.97\theta^{-.5}}$$

The logarithm and its derivative are:

$$l(\theta) = -5\ln\theta - 488.97\theta^{-.5}$$
$$l'(\theta) = -5\theta^{-1} + 244.485\theta^{-1.5}$$

Setting the derivative equal to zero yields

$$\hat{\theta} = (244.485/5)^2 = 2391$$

This answers the question. ▲

Question 17 *You are using an ARMA(1,1) model to represent a time series of* 100 *observations. You have determined:*

$$\begin{cases} \hat{y}_{100}(1) & = & 197.0 \\ \hat{\sigma}_{\varepsilon}^2 & = & 1.0 \end{cases}$$

Later, you observe that y_{101} *is* 188.0. *Determine the updated estimate* σ_{ε}^2.

Answer The estimated variance of the forecast errors is the sum of the squares of the error terms divided by $T - p - q$. In this case, after 100 observations, the sum of the squares of the error terms must equal 98, because the sum divided by $(100 - 1 - 1)$, or 98, is 1.0.

The 101st observation introduces a new error term equal to $188 - 197 = -9$. The square of this term is 81. Adding 81 to the previous sum of 98 gives a new total of 179. Dividing 179 by $101 - 1 - 1 = 99$ gives a new estimated variance of 1.8. ▲

Question 18 *You are given: (1) An individual automobile insured has annual claim frequencies that follow a Poisson distribution with mean* λ. *(2) An actuary's prior distribution for the parameter* λ *has probability density function*

$$\pi(\lambda) = (0.5)\, 5e^{-5\lambda} + (0.5)\frac{1}{5}e^{-\lambda/5}$$

(3) In the first policy year, no claims were observed for the insured. Determine the expected number of claims in the second policy year.

Answer The posterior distribution is

$$\pi(\lambda \mid 0) \propto e^{-\lambda}\left[(.5)\, 5e^{-5\lambda} + (.5)\,.2e^{-.2\lambda}\right]$$
$$= 2.5e^{-6\lambda} + .1e^{-1.2\lambda}$$

The normalizing constant can be obtained from

$$\int_0^\infty \left[2.5e^{-6\lambda} + .1e^{-1.2\lambda}\right] d\lambda = .5$$

and therefore the exact posterior density is $\pi(\lambda \mid 0) = 5e^{-6\lambda} + .2e^{-1.2\lambda}$.

The expected number of claims in the next year is the posterior mean,

$$E\left(\Lambda \mid 0\right) = \int_0^\infty \lambda \left[5e^{-6\lambda} + .2e^{-1.2\lambda}\right] d\lambda$$

$$= \frac{5}{36} + \frac{5}{36} = \frac{5}{18} = .278$$

This answers the question. ▲

Question 19 *During a one-year period, the number of accidents per day was distributed as follows:*

Accidents	0	1	2	3	4	5
Days	209	111	33	7	3	2

You use a chi-square test to measure the fit of a Poisson distribution with mean 0.60. The minimum expected number of observations in any group should be 5. The maximum possible number of groups should be used. Determine the chi-square statistic.

Answer There are 365 observations, so the expected count for k accidents is

$$365 p_k = 365 \frac{e^{-.6}(.6)^k}{k!}$$

which produces the following table:

Accidents	Observed	Expected	Chi-square
0	209	200.32	0.38
1	111	120.19	0.70
2	33	36.06	0.26
3	7	7.21	1.51
4	3	1.08	
5	2	0.14	

This answers the question. ▲

Question 21 *Twenty independent loss ratios $Y_1, Y_2, \ldots, Y_{20}$ are described by the model*

$$Y_t = \alpha + \varepsilon_t$$

where:

$$Var(\varepsilon_t) = 0.4, t = 1, 2, \ldots, 8$$
$$Var(\varepsilon_t) = 0.6, t = 9, 10, \ldots, 20$$

You are given:

$$\overline{Y}_1 = \tfrac{1}{8}(Y_1 + Y_2 + \cdots + Y_8)$$
$$\overline{Y}_2 = \tfrac{1}{12}(Y_9 + Y_{10} + \cdots + Y_{20})$$

Determine the weighted least squares estimator of α in terms of $\overline{Y}_1$ and $\overline{Y}_2$.

Answer We need

$$S(\alpha) = \sum_{t=1}^{8} \left(\frac{Y_t - \alpha}{\sqrt{0.4}}\right)^2 + \sum_{t=9}^{20} \left(\frac{Y_t - \alpha}{\sqrt{0.6}}\right)^2$$

to be a minimum. Setting the derivative equal to zero produces the equation

$$S'(\alpha) = \frac{1}{.4}\sum_{t=1}^{8} 2(Y_t - \alpha) + \frac{1}{.6}\sum_{t=9}^{20} 2(Y_t - \alpha) = 0$$

Multiplying by 0.6 produces the equation

$$0 = 3\left(8\overline{Y}_1 - 8\alpha\right) + 2\left(12\overline{Y}_2 - 12\alpha\right)$$
$$0 = 24\overline{Y}_1 + 24\overline{Y}_2 - 48\alpha$$
$$\alpha = .5\overline{Y}_1 + .5\overline{Y}_2$$

This answers the question. ▲

Question 22 *For a mortality study, you are given: (1) Ten adults were observed beginning at age 50. (2) Four deaths were recorded during the study at ages 52, 55, 58 and 60. The six survivors exited the study at age 60. (3) H_0 is a hypothesized cumulative hazard function with values as follows:*

$H_0(50) = 0.270$	$H_0(51) = 0.280$	$H_0(52) = 0.290$	$H_0(53) = 0.310$
$H_0(54) = 0.330$	$H_0(55) = 0.350$	$H_0(56) = 0.370$	$H_0(57) = 0.390$
$H_0(58) = 0.410$	$H_0(59) = 0.435$	$H_0(60) = 0.465$	

Determine the result of the one-sample log-rank test used to test whether the true cumulative hazard function differs from H_0. The possible answers are

(A)   *Reject at the 0.005 significance level.*

(B)   *Reject at the 0.01 significance level, but not at the 0.005 level.*

(C)   *Reject at the 0.025 significance level, but not at the 0.01 level.*

(D)   *Reject at the 0.05 significance level, but not at the 0.025 level.*

(E)   *Do not reject at the 0.05 significance level.*

Answer We are given that $O = 4$ and

$$E = (.29 - .27) + (.35 - .27) + (.41 - .27) + 7(.465 - .27)$$
$$= 1.605$$

Therefore, the chi-square statistic is

$$(4 - 1.605)^2 / 1.605 = 3.57$$

Hence the 0.05 level of significance is 3.84. So the answer is (E). ▲

Question 24 *Your claims manager has asserted that a procedural change in the claims department implemented on January 1, 1997 immediately reduced claim severity by* 20 *percent. You use a multiple regression model to test this assertion. For the dependent variable, Y, you calculate the average claim costs on closed claims by year during 1990-99. You define the variable X as the year. You also define a variable D as:*

$$D = \begin{cases} 0 & \text{for years 1996 and prior} \\ 1 & \text{for years 1997 and later} \end{cases}$$

Assuming a lognormal error component and constant inflation over the entire period, which of the following models would be used to test the assertion? The possible answers are

(A) $Y = \alpha_1^D \beta_1^X \varepsilon$

(B) $Y = \alpha_1 \alpha_2^D \beta_1^X \varepsilon$

(C) $Y = \alpha_1 \beta_1^X \beta_2^{XD} \varepsilon$

(D) $Y = \alpha_1 \alpha_2^X \beta_1^X \beta_2^{XD} \varepsilon$

(E) $Y = \alpha_1 \alpha_2^D X^{\beta_1} \varepsilon$

Answer With a lognormal error component, the linear model should be for the logarithm of the observation. A model that conforms to the description is

$$\ln Y = \alpha_1^* + \alpha_2^* D + \beta_1^* X + \varepsilon^*$$

Exponentiating both sides yields

$$Y = e^{\alpha_1^*} e^{\alpha_2^* D} e^{\beta_1^* X} e^{\varepsilon^*}$$

and then defining an unstarred quantity as its started version exponentiated, we have

$$Y = \alpha_1 \alpha_2^D \beta_1^X \varepsilon$$

Note that when D is 1, the value of Y is multiplied by α_2 and so the hypothesis to test is if this value is equal to 0.8. ▲

Question 28 *Two eight-sided dice, A and B, are used to determine the number of claims for an insured. The faces of each die are marked with either 0 or 1, representing the number of claims for that insured for the year.*

Die	Pr(Claims=0)	Pr(Claims=1)
A	$\frac{1}{4}$	$\frac{3}{4}$
B	$\frac{3}{4}$	$\frac{1}{4}$

Two spinners, X and Y, are used to determine claim cost. Spinner X has two areas marked 12 *and* c. *Spinner Y has only one area marked* 12.

Spinner	Pr(Cost=12)	Pr(Cost=c)
X	$\frac{1}{2}$	$\frac{1}{2}$
Y	1	0

To determine the losses for the year, a die is randomly selected from A and B and rolled. If a claim occurs, a spinner is randomly selected from X and Y and spun. For subsequent years, the same die and spinner are used to determine losses. Losses for the first year are 12. *Based upon the results of the first year, you determine that the expected losses for the second year are* 10. *Calculate* c.

Answer Let *EV* stand for "expected value" and *priorProb* for "prior probability" and *postProb* for "posterior probability."
    Then

Die/Spinner	priorProb	Probability of getting a 12	postProb
AX	$\frac{1}{4}$	$\frac{3}{4} \times \frac{1}{2} = \frac{3}{8}$	$\frac{1}{4}$
AY	$\frac{1}{4}$	$\frac{3}{4} \times 1 = \frac{3}{4}$	$\frac{1}{2}$
BX	$\frac{1}{4}$	$\frac{1}{4} \times \frac{1}{2} = \frac{1}{8}$	$\frac{1}{12}$
BY	$\frac{1}{4}$	$\frac{1}{4} \times 1 = \frac{1}{4}$	$\frac{1}{6}$
Total	1	$\frac{3}{2}$	1

and

Die/Spinner	EV	postProb	EV $\times$ postProb
AX	$\frac{3}{4} \times \frac{1}{2} \times (12+c)$	$\frac{1}{4}$	$1.125 + \frac{3}{32}c$
AY	$\frac{3}{4} \times (12)$	$\frac{1}{2}$	4.5
BX	$\frac{1}{4} \times \frac{1}{2} \times (12+c)$	$\frac{1}{12}$	$0.125 + \frac{1}{96}c$
BY	$\frac{1}{4} \times (12)$	$\frac{1}{6}$	0.5
Total		1	$6.25 + \frac{10}{96}c$

Since the expected value is 10 and $6.25 + \frac{10}{96}c = 10$, we have $c = 36$. ▲

Question 31 *For a study in which you are performing a proportional hazards regression using the Cox model, you are given that*

$$h(t|Z) = h_0(t)\exp\left(\beta^t Z\right)$$

and that the covariate vectors for the three individuals studied, in the order in which they die, are as follows:

$$Z_1 = \begin{pmatrix} 1 \\ 0 \\ 0 \end{pmatrix} \quad Z_2 = \begin{pmatrix} 0 \\ 1 \\ 0 \end{pmatrix} \quad Z_3 = \begin{pmatrix} 0 \\ 0 \\ 1 \end{pmatrix}$$

Determine the partial likelihood.

Answer By definition, we have

$$L(\hat{a}) = \left(\frac{e^{\beta_1}}{e^{\beta_1} + e^{\beta_2} + e^{\beta_3}}\right)\left(\frac{e^{\beta_2}}{e^{\beta_2} + e^{\beta_3}}\right)\left(\frac{e^{\beta_3}}{e^{\beta_3}}\right)$$

This answers the question. ▲

Question 34 *You are given the following claims settlement activity for a book of automobile claims as of the end of 1999:*

Year Reported/Year Settled	1997	1998	1999
1997	Unknown	3	1
1998		5	2
1999			4

and

$$L = (Year\ Settled - Year\ Reported)$$

is a random variable describing the time lag in settling a claim. The probability function of L is

$$f_L(l) = (1-p)\,p^l, \text{ for } l = 0,1,2,\ldots$$

Determine the maximum likelihood estimate of the parameter p.

Answer The observations are right and left truncated and the truncation depends upon the report year. For report year 1997 only claims settled at durations 1 and 2 can be observed, so the denominator must be the sum of those two probabilities. For 1998, only durations 0 and 1 can be observed and for 1999 only duration 0 can be observed. Calculation of the denominator probabilities is summarized below.

Probabilities

Year Reported	Settled in 1998	Settled in 1999	Sum (Denominator)
1997	$(1-p)\,p$	$(1-p)\,p^2$	$(1-p)\,p\,(1+p)$
1998	$(1-p)$	$(1-p)\,p$	$(1-p)(1+p)$
1999		$(1-p)$	$(1-p)$

The likelihood function is

$$L(p) = ABCDE = \frac{p^3}{(1+p)^{11}}$$

where

$$A = \left(\frac{(1-p)\,p}{(1-p)\,p\,(1+p)}\right)^3 = \left(\frac{1}{1+p}\right)^3$$

$$B = \left(\frac{(1-p)\,p^2}{(1-p)\,p\,(1+p)}\right)^1 = \left(\frac{p}{1+p}\right)^1$$

$$C = \left(\frac{(1-p)}{(1-p)(1+p)}\right)^5 = \left(\frac{1}{1+p}\right)^5$$

$$D = \left(\frac{(1-p)\,p}{(1-p)(1+p)}\right)^2 = \left(\frac{p}{1+p}\right)^2$$

$$E = \left(\frac{1-p}{1-p}\right)^4 = 1$$

The loglikelihood is therefore

$$l(p) = 3\ln(p) - 11\ln(1+p)$$

Taking the derivative with respect to p, we obtain the equation to solve:

$$\frac{3}{p} - \frac{11}{(1+p)} = 0$$

Therefore, the solution is $\hat{p} = \frac{3}{8}$. ▲

Question 40 *For a two-variable regression based on seven observations, you are given:*

$$(1) \quad \sum(X_i - \overline{X})^2 = 2000$$
$$(2) \quad \sum\hat{\varepsilon}_i^2 = 967$$

Calculate $s_{\hat{\beta}}$, the standard error of $\hat{\beta}$.

Answer Since

$$s^2 = \frac{\sum\hat{\varepsilon}_i^2}{N-2} = \frac{967}{5} = 193.4$$

$$s_{\hat{\beta}}^2 = \frac{s^2}{\sum x_i^2} = \frac{193.4}{2000} = 0.0967,$$

it follows that $s_{\hat{\beta}} = 0.31$. ▲

2.9 SOA Courses 5 through 8

Becoming a Fellow of the Society of Actuaries is similar to becoming a member of other professions such as doctors and lawyers. The process is arduous. In the case of the Society of Actuaries, eight examinations must be passed. While Courses 1 through 4 are of foundational nature, Courses 5 through 8 usually require work experience as well as theoretical knowledge. The question-and-answer sections shed some light on the importance of these courses in the life of working actuaries. Passing the examinations in the SOA Courses 1 through 5 is the first formal step in becoming an actuary. If you have passed these five examinations, you become an *Associate* of the Society of Actuaries and are entitled to put the title ASA after your name. For many members of the Society, becoming an Associate is also the last step because the work they do does not require them to pass additional examinations. The term "Career Associates" has been coined informally for those who stop writing examinations at this point in their careers. We illustrate the ideas and techniques making up the toolbox of full-fledged actuaries, by reproducing the goals and course descriptions of Courses 5 through 8, as given in Fall 2002 and Spring 2003 Basic Education Catalogs of the Society of Actuaries (*see* References 3 and 4, Appendix F).

SOA Course 5—Application of Basic Actuarial Principles

According to the SOA syllabus (*see* Reference 3, Appendix F), "this course develops the candidate's knowledge of basic actuarial principles applicable to a variety of financial security systems: life, health, property and casualty insurance, annuities, and retirement systems. The candidate will be required to understand the purpose of these systems, the design and development of financial security products, the concepts of anti-selection and risk classification factors, and the effects of regulation and taxation on these issues. The course will develop the candidate's knowledge of principles and practices applicable to the determination of premiums and rates and the valuation and funding of these financial security systems." The topics covered in this course are divided into the following topic areas:

- ▶ *Basic Principles of Design.* Here it is expected that you can explain and deal with problems of financial insecurity, product development, and methods of distribution.
- ▶ *Basic Principles of Risk Classification.* Here you are expected to classify risks involved in life insurance, health insurance, retirement plans, property and casualty insurance, and in non-traditional areas of insurance such as warranty. Here you are expected to be able to evaluate the risk classification factors and be able to carry out a cost/benefit analysis.
- ▶ *Basic Principles of Pricing/Ratemaking/Funding.* Here you are expected to be able to describe the objectives of various coverages, evaluate the assumptions underlying pricing, and describe the major pricing and funding techniques and methods used in life insurance, health insurance, retirement plans, and property and casualty insurance. You're also expected to be able to develop different types of profit/surplus measure and describe methods for evaluating pricing.
- ▶ *Basic Principles of Valuation.* Here you are expected to be able to describe valuations and the different purposes for performing a valuation, and be able to determine the actuarial value resulting from applying the methodology. You are also expected to be able to interpret the results of the valuation.

SOA Course 6—Finance and Investment

According to the SOA syllabus (*see* Reference 4, Appendix F), "this course extends the candidate's knowledge of basic actuarial principles in the fields of investments and asset management. Candidates completing this course will have developed some expertise in the areas of capital markets, investment vehicles, derivatives-applications, principles of portfolio management and asset-liability management."

SOA Course 7—Applied Modeling

Course 7 is a seminar-type course and laptops are required. According to the SOA syllabus (*see* Reference 4, Appendix F), "this course introduces the candidate to the practical considerations of modeling through an intensive seminar using a case study format.... The interactive approach of the seminar will require candidates to draw upon knowledge from the basic courses and learn applied modeling skills in a hands-on environment. The seminar also emphasizes communication skills, teamwork and the synthesis of subjects in an applied setting."

Course 7 draws on the students' experience in modeling, problem solving and communication and have sufficient technical knowledge of a limited number of models to be able to benefit from the course. It is expected that students are familiar with the idea of an actuarial model and have a broad understanding of their use in actuarial practice involving survival models, credibility models, risk theory models, ruin theory models, option pricing models, cash flow and cash flow testing models and non-traditional models. Students must be able to apply appropriate models to solve business problems and be able to analyze and understand the results of the modeling process. Moreover, they must be able to effectively explain their work and results to others.

SOA Course 8—Advanced Specialized Actuarial Practice

As explained in the SOA syllabus (*see* Reference 3, Appendix F), this course is divided into several options: finance (corporate, capital management, financial risk management, financial strategies); health, group life and managed care (plan design, data and cost analysis and rating, financial management, administration and delivery systems); individual insurance (marketing of individual life insurance and annuity products, pricing, valuation and financial statements, product development and design); investments (portfolio management, option pricing techniques, asset-liability management), retirement benefits. Students are required to choose one of these areas of specialization. In all areas, students are expected to be familiar with the basic ideas and techniques, to develop models and strategies, and evaluate and communicate the consequences of their choices.

2.10 | CAS Courses 5 through 9

As in the SOA case, you can become an *Associate* of the Casualty Actuarial Society by passing the first five CAS examinations. You can then use the letters ACAS after your name. After passing the remaining four examinations in Courses 6 through 9 (CAS), you become a *Fellow* and use the letters FCAS. Starting in 2003, you will also have to write a CAS-only examination in Course 3. The syllabus of Courses 5 through 9 is again based on the work experience of the candidates. For many members of the Society, becoming an Associate is also the last step because the work they do does not require them to pass additional examinations. We illustrate the ideas and techniques making up the toolbox of full-fledged actuaries by summarizing the goals and course descriptions of Courses 5 through 8, as given on the 2003 website of the Casualty Actuarial Society. The descriptions will give you a taste of the kind of knowledge and experienced needed as a casualty actuary. You should consult the CAS website for complete and up-to-date statements of the learning objects and examination requirements for these courses. The results of the 2002 CAS survey on professional skills (*see* Reference 7, Appendix F) give some idea of the dynamics of the impending curricular change.

CAS Course 5—Introduction to Property and Casualty Insurance and Ratemaking

Course 5 deals in part with the legal and commercial nature of insurance policies and coverage. Actuaries should be able to understand the fine print on an insurance policy. They should have "an understanding of the nature of the coverages provided and the exposure bases used in the respective lines of insurance." They should understand the connection between coverage and pricing and be able to interpret the conditions, exclusions, and limitations of P/C policies. To do so, they must be familiar with manual excerpts and must study illustrative parts of relevant manuals dealing with forms, coverages, and rating process.

The course also covers insurance company operations including company organization, marketing and distributions systems, underwriting, and claims. In addition, P/C actuaries need to have a thorough understanding of the underwriting function including purpose, principles, and activities. They should also know how to settle claims based on policy provisions and have an understanding of the impact of settlements on overall loss levels.

P/C actuaries must understand the basic principles of ratemaking and be able to analyze data, select appropriate techniques, and have the tools to solve numerical problems. They should be able to compare the relative advantages and disadvantages of different procedures. In more general terms, P/C actuaries must be able to relate changes in the economic environment to the pricing of insurance.

CAS Course 6—Reserving, Insurance Accounting Principles, and Reinsurance

The components of Course 6 are statements of principles and standards of practice of insurance, dynamic financial analysis (DFA), expense analysis, published financial information, and reinsurance. In particular, they should be able to establish and review actuarial reserves, select and evaluate loss reserving methods for known claims and for claims incurred but not yet reported (IBNR). Property and casualty actuaries must have a general knowledge of insurance accounting. They must be able to explain the differences between the different accounting methods and be able to interpret and evaluate numerical data from the reports. In addition, they must understand the ideas and techniques involved in insurance companies insuring other insurance companies. The activity is known as *reinsurance*. They must be familiar with different types of reinsurance, the purposes of reinsurance, and how it is marketed and underwritten. They must understand how concepts such as pricing and reserving are adapted to apply to reinsurance.

CAS Course 7 (US)—Annual Statement, Taxation, and Regulation

Course 7 is country-specific since it involves reporting principles, taxation, and other regulations that differ from country to country. At this point, the CAS course is offered in two flavors, American and Canadian. The course consists of two main parts: insurance law and regulations, and accounting. Property and casualty actuaries must understand different aspects of insurance regulation and laws, markets, coverages and private and governmental programs. They must be able to assess how these regulations and laws impact on property/casualty coverages, ratemaking, and pricing. In the United States, this includes the tort law, statutory insurance and governmental programs such as social security and Medicare, catastrophes, and workers' compensation. They must also understand the role of antitrust law as it pertains to insurance regulation. They must understand the impact of government regulations on ratemaking, profitability, risk classification, and the availability of insurance. The course also covers the regulation for solvency, the IRIS [Insurance regulatory information systems] financial ratio test and guaranty fund mechanisms set up by the various states. The US version of the course covers the aspects of statutory and GAAP [generally accepted accounting principles] insurance accounting and taxation as they affect reserving and statutory reporting, and insurance company audits. The course assumes a working knowledge of general accounting such as that gained from Course 6.

CAS Course 7 (Canada)—Annual Statement, Taxation, and Regulation

The Canadian version of Course 7 includes a comprehensive presentation of Canadian tort law in the perspective of the insurance business in Canada. The course focuses on insurance regulation and insurance contract law and includes an overview of federal and provincial insurance programs. It also covers finance and solvency issues. It includes insurance accounting and its relevant laws and regulations, solvency monitoring systems such as the Dynamic Capital Adequacy Testing of the Canadian Institute of Actuaries. The course also includes sections on background law and insurance, the regulations of insurance in Canada, insurance as an essential service, and federal and provincial government plans such as the principles and ideas underlying Canadian employment insurance and the Canadian pension programs. The course includes material regarding environmental liabilities in the United States and on Canadian earthquake guidelines. It also covers Canadian provincial health plans, the regulatory environment surrounding US workers compensation. Canadian P/C actuaries must also understand Canadian automobile insurance programs including no-fault concepts and residual market requirements, and provincial guaranty funds. The finance and solvency section of the course deals with finance, taxation, and solvency tests. Canadian P/C actuaries need to be familiar with the concept of an *Annual Return*. This includes recent guidelines from OSFI [Office of the Superintendent of Financial Institutions] and the provincial regulatory bodies. A thorough knowledge of the GAAP [generally accepted accounting principles] is also required. The course covers such solvency monitoring systems such as the minimum capital test, risk-based capital requirements and the DCAT [dynamic capital adequacy testing] method of the Canadian Institute of Actuaries.

CAS Course 8—Investments and Financial Analysis

Course 8 deals with a broad array of finance, investment, and financial risk management topics. Its two main parts are financial theory and financial analysis. The course builds on the topics covered in Course 2. It also assumes knowledge about liability and reserve risk from Course 6, some knowledge of underwriting from Course 5, and knowledge of models and modeling from Courses 3 and 4.

Financial theory deals with investments with an emphasis on the cash flow characteristics, value, and risks inherent in various financial instruments. In particular, it deals with financial instruments and markets, portfolio theory, equilibrium in capital markets, CAPM [capital asset pricing model], index models and arbitrage pricing. Once of the key concepts covered is that of market efficiency. In addition, the financial theory part of the course covers fixed income securities, options, futures, and swaps, and international securities.

The financial analysis part of the course emphasizes measuring and managing the financial risk and overall value of an insurance company. It includes asset liability management and factors that affect the price sensitivity of fixed income securities. It also deals with various ways in which a portfolio manager can manage the interest rate and cash flow risk in a portfolio.

CAS Course 9—Advanced Ratemaking, Rate of Return, and Individual Risk Rating Plans

Course 9 deals with "the types of practical problems that a fully qualified actuary working in ratemaking should be able to solve." The techniques covered are divided into four sections: classification ratemaking topics; excess and deductible rating; rate of return; and the loading for risk. The excess and deductible section deals with methods of estimating losses within layers of coverage. The rate-of-return part of the course "explores the relationship between insurance concepts (such as underwriting profits, premium-to-surplus ratios, and investment income) and financial concepts (such as interest rates, inflation rates, cost of capital, and risk premiums)." The loading-for-risk part of the course concentrates on the fortuitous nature of insurance claims, the fact that the loading for profit in rates may not be realized. Individual risk rating is one of the important functions performed by an actuary in the rating of individual risks. The earlier courses dealt mainly with group insurance and classification risk rating. Course 9, on the other hand, is meant to enable P/C actuaries to design and manage individual risk rating systems. The three key concepts involved are experience rating (using individual risk experience to adjust rates), retrospective rating (using individual risk experience to adjust premiums after the completion of policies), and excess and deductible rating, (excluding portions of the individual risk experience from insurance coverage, and prospectively reducing rates). It is assumed that students "have a good working knowledge of credibility, loss limitation, and rate modification concepts as they apply to prospective and retrospective rating. In addition, they will be expected to have knowledge of loss distribution, insurance charge, and excess loss charge concepts as they apply to loss retention programs."

2.11 Other Courses

As is the case for other professions such as engineering, accounting, law, dentistry, and medicine, and so on, members of the profession are bound by explicit rules of conduct and a profession-specific code of ethics. Actuaries are no exception. Their educational systems usually include a professional development component similar to the PD [professional development] requirement of the Society of Actuaries. In all accreditation systems, be they university-based, profession-based, or a hybrid of the two, the professional societies of the respective countries examine candidates in this area before admitting them to the profession. The specific rules for accreditation are country-specific and are explained in detail on the websites listed at the end of the book.

Chapter 3

ACTUARIAL JOBS

Actuaries primarily provide actuarial services. Given their role in society world-wide, it is not surprising that a web search under *actuarial services* may produce over 200,000 hits. Any such search shows that actuarial employers come in all sizes, large and small. Actuaries work not only in consulting firms and insurance companies, they also work in government departments, in the human resource offices of most major companies, and in small firms specializing in a variety of actuarial tasks.

It is difficult to present an accurate snapshot of the state of the actuarial world at any given moment in time. Globalization of economies and changes in financial regulations and structures often blur the distinction between the actuarial and non-actuarial role of many companies. In this text, therefore, we concentrate on representative companies, large and diversified enough to illustrate the spectrum of actuarial careers. You will find basic information about some of the top employers of actuaries. The list does not include alternative forms of employment in business, banking, teaching, human resource management, administration, government, and so on. For more detailed information on actuarial careers elsewhere, we refer to the websites listed in Appendix D at the end of this book. The profiles vary from company to company to illustrate different aspects of actuarial employment.

We begin by answering some of the basic questions you may have about getting started with your actuarial career.

3.1 | Landing Your First Job

Q What advice and tips would you give, from an employer's point of view, to students applying for an actuarial position in your company, based on your participation in recruiting activities?

Answer Do more "not-actuarial-related activities," be yourself in interviews, don't try to do too much in interviews, show you are more than just a worker, but also a great person, etc. (There is more to life than grades and exams!)

Answer Write exams as quickly as possible.

Answer Be honest. Be yourself. Good grades are important, but companies look more and more for extracurricular activities. It's better to have a B but be involved in your actuarial association while doing volunteer work for a youth center than devoting your life to your A+. They want to know you're going to be efficient but also fun to work with, that you'll have something different to bring the company. Participate in conventions, wine and cheese events, and meetings with future employers. Even if you don't give them your CV, they'll probably remember you.

Answer To be energetic, to have a good team spirit, to be able to have activities while studying. Internship work is also well recognized. Good grades (not necessarily excellent grades) are also required.

Answer Have at least two exams (for full-time). Have someone proofread your letter and résumé, or at least use spell check—people who misspell "actuarial" are eliminated. I should have the idea that you are writing to me, or my company. State that you know this is a casualty company, and use CAS exams (not SOA). If you are a really good candidate (exams, A+ marks in an actuarial program, work experience whether in insurance or not, a great personality), you don't need to apply to a vast number of companies to get a job—you have time to customize. If you aren't a really good candidate, then you have to customize to get my attention. Keep the cover letter to one page. Take interview training. Be able to answer the standard questions. The more you talk, the more I will feel that I know you when you leave the interview. Look at our website before you apply to us (or for interns, before your interview)—almost everyone does. Work at something—it gives of us more to talk about.

Answer Graduate from university with as many exams as possible. Get relevant work experience through internships and summer jobs. Actuarial employers are generally located in large cities so be prepared to move to one if you don't already live in one. Make sure you develop both actuarial, computer, and communications skills. Pick the job that you'll think you'll enjoy the most—don't necessarily pick the one offering the highest salary.

Answer Always apply for job openings. Don't be afraid of a company's reputation. Grades are not everything.

Answer Take the intership opportunities to taste various forms of work and working environments. Try to see where you feel better. Then come to see me and tell me why you want to come and work for me.

Answer Résumé must be free of grammatical errors. Good marks are the most important factor—they need not all be 90+, but we do not interview very many students with an average below 70. Come to the interview well-dressed and professional. Look like you would be ready to work today. If in doubt about "business casual," take it up a notch. Listen to what the interviewer is saying. You will be evaluated for listening skills almost as much as speaking skills.

Answer Be genuine, without bragging. I've found in the past that students walk in like they own the world and think they know everything because they've had A+ in school. It's usually a very humbling experience when you start work and realize that you know next to nothing.

Answer Keep the university grades above average. Oftentimes, the only way to tell candidates apart (prior to an interview) is to look at the grades. Try and get some experience working for an insurance company while in university. Write a few exams to show your dedication and ability to write them.

Answer Have your résumé done professionally. Make sure there are no mistakes in it. Be prepared to relocate.

Answer Pass as many SOA exams as possible during university year. It is easier to study at school than at work even if the employer gives us studying days. Be confident. Show interest. Do a research on the company fields of practice, goals, successes, etc.

3.2 Moving Up the Ladder

Q How fast can an actuary expect to climb the ladder in your company? Illustrate your answer and compare it with examples from other companies you may be familiar with.

Answer There is not a typical path but I know that by the nature of the work, people in Asset Consulting become consultants and responsible for clients after fewer years than in Retirements or Health and Welfare Line of Business for example. I can't compare with other companies.

Answer It depends on the number of exams and how devoted the actuary is.

Answer You start as an analyst and become a consultant after 6 to 10 years. Then you can become a senior consultant after another 5 years.

Answer In insurance companies, in order to climb the ladder, there must be "pace" up in the ladder. That is what I have observed, therefore the pace will be different for everyone. In consulting companies, such as Mercer and Hewitt, I have observed a more "parallel" way of going up. If one is good at what one does, then the possibility of moving up is there. Everyone seems to be given the same chance and I believe that climbing the ladder in the consulting business can be faster than in an insurance company where the years of dedication and work are usually rewarded.

Answer It's all based on exam performance. It is possible to pass exams too quickly, i.e., get your Fellowship without much work experience (under 5 years). A company would probably be reluctant to promote someone to management in that case. But it's still never a bad thing to pass exams, even if your work experience isn't commensurate with your exam success. That's a better spot to be in than to have a lot of experience, but struggling with exams.

Answer Will generally depend on quality of work, understanding issues, learning quickly, time and effort, ability to solve problems (find solutions), good communication skills.

Answer Depends on the number of hours willing to put in, rapidity in passing exams and the efficacy of your work.

Answer Being an actuary is irrelevant to any ladder climbing I could do here. On the contrary, I should become more and more of a HR generalist if I wanted to go much higher.

Answer From new hire to Assistant Vice-President (officer) in ten years. Not at all uncommon in six or seven, depending on exam progress.

Answer In Canada, for P/C companies, most actuaries are at least managers by the time they are 30, with a fair number even being VP. Again, it all depends on the individuals, and their willingness to take risks. One may have to move to where the opportunities are, for example. It also depends on the size of the company. Opportunities at my prior company were limited, since there was a fair number of Fellows employed there. Still, with my departure, one opportunity was created. Sometimes, it is also a matter of being at the right place, at the right time.

3.3 Salaries and Benefits

Q How do you negotiate your salary, and have you always been satisfied with the salary and benefit packages you have had? Illustrate your answer from your knowledge of the practice of specific companies.

Answer I never really negotiated salaries and things like that since Towers Perrin has their own way to give salary increases and so on. So far I have felt satisfied with the way I was treated "money-wise," but I would not be afraid to discuss it with my supervisor if that was the case.

Answer Companies offer packages they believe are fair, and apply internal equity. Rarely are they negotiated.

Answer It is important to be aware of the market when negotiating. I used the D. W. Simpson website. It's a recruiting company specialized in actuaries. They often run surveys on salary by exams and years of experience and post their results on their site. Also, it is important to know the company you're applying for. The size of the company and the field will influence the salary (insurance versus consultation, small company from Montreal versus international company). It is important to remember that not only the salary counts. The other benefits (health plan, stock purchase program, bonuses, study days), responsibilities, chances of promotions and environment, for example, should all be considered. So far, I've always been satisfied with my salary and benefits.

Answer Not negotiated. I accepted an offer and did not really test the waters with other companies.

Answer It is hard to negotiate salary and I don't think my employer offers a competitive package. They actually tell us that we are paid on "average" and not on the high salaries. Bonuses are great though (from 15% at hire to 25% for consultants).

Answer As an intern in companies, I am not in a position to negotiate my salary, but I do know what to expect. Depending on the number of exams passed, the number of years in school and the work experience, one's salary can vary. For my part, I have been very satisfied with the salary I have received.

Answer I generally don't negotiate mine unless I'm changing companies. Then I negotiate to the point to which the salaries are roughly consistent across offers and I am picking my job based on the job, rather than the compensation. I used to work in the United States and they pay more. But they sometimes low-ball Canadian students who don't know the US actuarial student market. I took a pay cut to return to Canada. So salary isn't the absolutely highest priority I have. I think that, given the lower cost of living in Canada, the standard of living of American and Canadian actuaries is about the same. I think that insurance companies always offer a very competitive benefits package—medical, retirement, cafeteria, fitness centre benefits that many other companies just don't have. Larger companies can afford to have more comprehensive benefits than smaller ones, I think.

Answer No real negotiation. Performance objectives are set in advance and salary is determined based on reaching those objectives. I have been satisfied with my salary most of the time.

Answer Tough to do because there is no one to compare to. Firms that hire many actuaries (consultants, insurers) have a good knowledge of the market, but others don't. Not that I have always been satisfied (who is?), but in this field, money isn't everything. You have to look at the quality of life also.

Answer A new student has little flexibility. A student hired from another company has more ability to set the salary. Usually the hiring company will want to pay no more than 10% or 20% above what you are currently getting (this will depend on how long you have been in your current job and how hot the company is to get you). Remember that although you are an attractive commodity, if you overprice yourself you had better be able to deliver superb value or risk a reputation as a bad deal.

Answer I start with the assumption that the company that I work for treats me equitably. If I discover that this is not the case, I don't hesitate to make a move. Until now I believe that only one of my employers (I have had three so far) was not paying me what I was worth, and I left him after nine months. Even though he wanted to correct the situation (with a 30% increase), I quit because my relationship with my employer is one of mutual trust.

Answer At the entry level and intermediate positions in my company, salaries are quite competitive with the market. Negotiation usually happens at the time of hire and future increases are based on merit, responsibility, level and performance. Since I work for a very large employer with many people at or around the same level, I believe salaries are fair and competitive. Unless you are a superstar, there are probably only minor salary negotiations. Of course if you are good and the competition actively recruits you, you will most likely receive a higher pay as an incentive.

Answer In my experience, I never had to negotiate much for my salary. Headhunters are usually a great help in negotiating on your behalf with a new company. I must admit I have also generally been satisfied with my salary and benefits, although other people out there may have better conditions. In turn, I'm probably doing better than other people out there. At the end of the day, what really counts is being happy with what you do and whom you work with.

Answer For positions that are non-managerial, companies usually have salary scales, which they usually follow. There is not a lot of room for negotiation. For managerial positions, salary is usually based on the experience and the skills of the candidate, as well as "current salary" as a basic input. This is done on a case-by-case basis, there is no general rule. At the executive level, a company will usually want a specific individual. Therefore, there is much more room for negotiation. There obviously is no general rule applicable in this case. Benefit packages are always very good.

Many employers have actuarial training programs for actuarial students. A typical training program will include assistance with exam preparation, rotational assignments to expose to the student to different aspects of actuarial work, seminars, purchase of study material, paid study time, and the reimbursement of the costs associated with the writing of the examinations. The survey illustrates the value of such training programs for career development.

Q Based on your employment experience, describe the support different companies give to students to prepare for actuarial examinations (study days, payment of examination fees, purchase of study material, etc.)

Answer Most employers who retain the services of actuaries generally provide full support.

Answer We have great support from the firm: three study days per hour of exams, all fees/books/study guides are paid, we can ask for a seminar support if needed, we have passing bonus in dollars and pay increase too that are very good when compared to the market. Lots of support from fellow workers too. So our workload is not as huge in the final weeks of studying.

Answer Most provide the same level of study days, three days per hour of exams, payment of fees and study material. Very few will pay for seminar.

Answer Usually the payment of examination fees, the study seminar support, purchase of study material and the study aids is mostly the same from company to company. What I find the most important is the amount of study days (or the respect of your study time). This is what varies the most. Sometimes, you have to work overtime and on weekends to be able to take your study day.... And it happens often that you can't even take your study day because you have a rush. This applies in the consultation field, it's not as true in the insurance field and doesn't apply at all if you work for the government. Instead, they should set a maximum numbers of working hours when you're studying.

Answer For interns, most companies offer the same studying benefits. We usually get six study days, and the day of the exam off, and while some companies pay for books, and the exams, others will only pay for the exam if you have successfully passed it.

Answer Study days are generally based on the number of hours for the exam (that is three days per hour up to 18 or 20 per session). Usually the exam fees are covered for the first (sometimes second) writing of an exam. After that they may be reimbursed upon passing of the exam only. The study materials and aids are generally purchased by the company.

Answer We have a good study-time program. About 15 to 18 days for exams 5 and over. They pay 100% of the material and exam fees unless you failed too many times (I don't know how many!). They don't pay for support seminar.

Answer I have found that most insurance companies are pretty similar in their student program. All pay the exam fees, study materials and give a certain amount of studying days off to the student. If the student fails a course on his second attempt, the studying time is shorter but the other advantages stay the same. Most companies will stop paying exam fees, study time and material after the third trial but will reimburse the fees if the exam is passed. Companies also offer bonuses upon successful completion of a course. I have found this to be very much the same in all companies, both in insurance and consulting companies.

Answer Exam support seems to be pretty universal across the different companies. All companies offer study time, payment of exam fees, materials, etc. Some might pay for exam seminars. The study probably won't vary from company to company too much, although whether you *actually* get to use all of your time is different. Insurance companies stress passing exams more so their students usually get their full study time allotment. Consulting companies usually work their students to the point that they lose out on some of their time or they have to try to cram it all at the end. If a company offers more study time, then they usually have a higher standard of required successful exam attempts to stay in their program.

Answer Generally: three study days per hour of examination paid by the company, examination fees paid by the company, study material and study aids paid by the company, study seminar may require approval. For unsuccessful candidates: for a third try no study days are provided (must take vacation days), only half of examination fees are paid and the other half is reimbursed if the exam is passed.

Answer The information below is based on my personal experience and what I have heard from actuaries in other companies. Most companies tend to give approximately 15 study days for a candidate's first attempt at a given exam. The variation between companies tends to come more for the second attempt, where I have seen anywhere from 7.5 to 12 days. Third and fourth attempts can be from 0 to 7.5 days. Most companies also tend to pay for the exam fees at least once and usually twice, if the candidate passes on the second attempt. Third or fourth attempts are often paid by the candidates themselves. Study materials are usually provided by the employer.

Answer Study days are a must, and you may have to negotiate for them if you join a small company. Sometimes fees are always paid no questions asked; sometimes they are paid upon a passing grade; sometimes they are paid for the first two tries, then 50% on the third try, then you're on your own, my friend. I've seen a wide variety depending on the companies.

Answer Consulting firms tend to give more support than insurance companies. Mercer seems to be the best of them all. The give paid study days and pay for all the necessary exam materials.

Answer Fifteen days study time, all fees paid for first attempt, partial for repeat, study aids paid for, company pays for seminar costs, travel for PD [professional development], etc.

Answer With respect to my company: study days are three days per hour of exam, cost of exam paid up front, study materials are paid up front. Students can keep their study aids (ACTEX, JAM, etc.), but the books must be returned to the study library; study seminars, up to 50% of the cost is paid up front, and the remainder is paid upon passing the exam.

Answer Most companies will give about fifteen paid study days per four hours exam, pay for successful completion of an exam, and buy the study material. Some companies will pay for part or all of a study seminar. Most companies will now penalize employees who consistently fail exams by reducing the number of study days provided upon the second and third failure of the same exam.

Answer Twelve to 16 study days. Exam fees, material and study aids reimbursed/paid up front for first two attempts. Salary increase upon successful completion of an examination. Salary increase when the ACAS and FCAS designations are attained. Partial reimbursement for out-of-town study seminars. Cash bonus if successful on first attempt.

3.4   Company Reputation

In this section, several respondents of the survey comment on the quality of the work environment of past and current employers.

Q **Are there companies with a good or bad reputation in the actuarial industry? Give some examples and explain why.**

Answer Mercer Benefits Consulting has a tremendously good reputation in the industry as a fair and rigorous training ground for younger professionals. Furthermore, their global presence provides valuable opportunities for transfers.

Answer I have found that at Towers Perrin people really enjoy being in the environment we have here. Lots of flexibility from everyone, working hours and holidays, senior people help the younger ones a lot and I really appreciate the support I am getting. Very few people have left the company since I have been here. I can't really speak about other companies.

Answer Consulting actuaries are known to work a lot, maybe too much. They depend on clients and want to do everything by yesterday because they think the clients will be more satisfied. But it's a challenging job. You get to work with clients and in teams, where interpersonal skills are very important. At the other end of the spectrum, actuaries working for the government is just other white collar workers. They may not always work a lot, may not have high salaries and may be do boring or repetitive job. But they will have more time to study and so will probably become Fellows faster.

Answer Consulting firms are recognized to make their employees work a lot but from my personal experience, the workload gets better after a few years.

Answer Most major companies are respected by the others. I think the major companies realize that all companies need to have a good reputation so that the industry itself does not suffer.

Insurance companies sell intangible goods—they sell promises. They promise to pay money when they're contractually obligated to. So if people start believing that insurance companies can't be trusted to keep their promises, then every company suffers.

Companies may have differences of opinion on what qualifies as ethical sales methods—policy replacements and churning. I'm reluctant to name any specific company with a bad reputation because my experience is that unethical behavior is often restricted to isolated cases of miscommunications or over-zealous field agents, rather than an unethical systemic problem knowingly practiced by the company.

Answer Yes and no. Each company has its own set of core values that are disseminated by more senior employees. These values will fit with some people, clients, etc. Hence, there will always be a need for different companies to fit everyone's needs.

Answer Reputations change over time. Big companies tend to attract good students and spend a lot of time training them. They cannot risk bad publicity. Small companies are often reputed to underbid.

Answer Consultants had a bad reputation of making you work very hard, but things have changed and now they are fighting over the new graduates. What's tough about consulting is that you are told what to do (not much room for your own input before you reach the strategic consultant level). I don't find that I work less since leaving the consulting business. It's just that now, my work is more focused and I have much more influence over the decision-makers.

Answer Mercer is outstanding. They hire only the best people so it creates a great work/learning environment. Who better to learn from than the best in their field?

Answer Manulife has a good reputation because we have a lot of actuaries and we are a big company with many opportunities.

Answer Consulting firms usually have the reputation of being "sweatshops." To a certain extent, consultants may put in more hours than insurance actuaries, for example, but the pay at higher levels is also more significant. Consulting actuaries work more than nine-to-five, but the hours can be flexible, and calling them sweatshops is probably an exaggeration.

Answer I'm not sure that there are companies with a good or bad reputation in the same way in which there are individuals with good or bad reputation. Not all actuaries are qualified to be managers. However, by virtue of their Fellowship, most actuaries do become managers. Whereas some of them are just not good managers, some others are actually bad managers. By this I mean that people will work for them for short periods (usually less than a year) and move on to another company. This results in the company having a revolving door (and a bad reputation), where actuaries leave every few months.

In order to find a company with good managers, students should to talk to other students, ideally ones who have just left a particular company, or ones who are currently working there, before deciding to change jobs. If you find yourself, each and every morning, not wanting to go to work, you probably have a bad manager, and I would advise changing companies. It may also be that an actuarial career is not for you, in which case you should change careers!

3.5 | Consulting or Insurance

Actuarial careers can be looked at in several ways. Consulting versus insurance is one of them. Here are some comments from respondents to the survey about the advantages of these career options:

Q What are the main advantages of working for (a) a consulting firm, (b) an insurance company?

Answer Consulting firms: flexible schedule, diversity of work, relations with clients (sometimes a downside), lots of learning opportunities and career path, meeting interesting people with different expertise and backgrounds.

Answer Consulting offers better long-term compensation conditions, and is more client-oriented.

Answer In an insurance company you will generally have shorter day of work. The work you have to do is for your company so you will be less squeezed with urgent deadlines. For me, the main advantage is that you will be working with people who have the same skills as you. If you have something to explain, it will be to someone that understands how actuarial mathematics works. The main advantage of a consulting firm is the contact with clients. It makes the work more different from one day to another. Also, you might be asked to develop new programs, thus you are constantly learning new stuff.

Answer Insurance companies: more stable work schedule, salary increase for each exam pass, more technical work at junior levels. Consulting firms: more communication involved such as meeting with clients, challenged by peers about current issues, etc.

Answer Insurance companies: job stability. The regular hours and more relaxed atmosphere. Working in an insurance company requires more technical skills so if someone prefers the technical work, that's their place!

Consulting firms: the interaction with clients, the variety of the work, and the chance to work in larger teams.

Answer Having worked in both, I believe that it boils down to who you are working for, the team, and what work is available. The work is often very similar.

Insurance companies: opportunity to do more types of projects than just loss reserves and rate filings. For those who will pass exams more slowly, probably better opportunities.

Consulting firms: tend to attract and retain those who are bright and willing to work a bit harder for the money (although those at companies can work as hard or harder). Can be more variety among types of projects (reserves, rates, types of clients). Generally higher standards, especially for documentation. Generally faster promotions for those who pass exams. Fewer administrative meetings. Projects tend to have definite end (unlike insurance companies, where projects can be put on the back burner forever).

Answer Insurance companies: more exam support, pass through exams quicker, get promoted more quickly as a result, rotation program in different areas of the company.

Consulting firms: higher initial salary and ultimately more rewarding if you can manage to pass your exams quickly.

Answer Consulting firm advantages: higher salary, exposure to a wider range of actuarial topics. Greater sense of ownership for your work as it affects the results of your company.

Consulting firm disadvantages: longer hours, more pressure, sometimes difficult to fit in study days.

Answer Insurance companies: stable workflow, job you keep for a long time. Consulting firms: stimulating work and younger colleagues!

Answer In a consulting firm you work longer hours, but you have the advantage of being your own boss. You work the hours that you would like to. In a consulting firm there is less of a hierarchical structure. Everyone who merits it will get to be a boss/consultant. There is more of an entrepreneurial spirit since you work directly with your clients and there is more of a personal profit incentive.

Answer As a student, you will touch more varied projects working for a consulting firm. There is also a lot less programming and data entry involved with consulting firm. However, the best consultants are the ones who understand how an insurance company works, and what are the challenges of running a company. The only way to learn this is to work for an insurance company. Once higher up in the organization, being involved with the day-to-day management of a company is also a rewarding experience, which consulting cannot offer.

Answer Insurance companies: lower workload (apparently), less pressure to produce (apparently), better knowledge of the business because of focus on one company/group of companies, more involvement in decision-making process, comfort derived from knowing the environment and the people. Consulting firms: higher income, higher standards, no inter-departmental politics ...

Answer Insurance companies: less stress, more flexible schedule, fewer work hours, social benefits more generous. Consulting firms: More challenging, possibility to have a promotion, higher salary, more diversified work.

CONSULTING FIRMS

In this appendix, we profile a number of typical actuarial consulting firms. Most of these companies provide employment opportunities that go far beyond the actuarial field. Unless otherwise stated, statistics and quotes are taken from the company websites.

AON CORPORATION

Headquarters
  Aon Center
  200 East Randolph
  Chicago, Illinois 60601
  Phone: (312) 381-1000
Internet: www.aon.com

Aon is a Fortune 500 company that is a world leader in risk management, retail, reinsurance and wholesale brokerage, claims management, specialty services, and human capital consulting services. The company has an employee base of 55,000 people working in 600 offices in more than 125 countries. In the United States, Aon has offices in all states.

Locations

Albania, Antigua (West Indies), Argentina, Aruba, Australia, Austria, Azerbaijan, Bahamas, Bahrain, Bangladesh, Barbados, Belgium, Belize, Bermuda, Bolivia, Bosnia, Botswana, Brazil, British Virgin Islands, Bulgaria, Canada, Cayman Islands, Chile, China, Colombia, Costa Rica, Croatia, Cyprus, Czech Republic,

Ecuador, Egypt, El Salvador, Estonia, Ethiopia, Fiji, Finland, France, Georgia, Germany, Ghana, Gibraltar, Greece, Grenada, Guadeloupe, Guam, Guatemala, Guernsey, Guyana, Haiti, Honduras, Hong Kong, Hungary, India, Indonesia, Ireland, Isle of Man, Israel, Italy, Jamaica, Japan, Kazakhstan, Kenya, Kosovo, Kuwait, Latvia, Lebanon, Lesotho, Lithuania, Luxembourg, Macedonia, Malawi, Malaysia, Malta, Mauritius, Mexico, Morocco, Mozambique, Myanmar, Namibia, Netherlands, Netherlands Antilles (Dutch Carribean), New Zealand, Nicaragua, Nigeria, Northern Ireland, Norway, Oman, Pakistan, Panama, Papua New Guinea, Paraguay, Peru, Philippines, Poland, Portugal, Puerto Rico, Romania, Russia, Saipan, Saudi Arabia, Singapore, Slovak Republic, Slovenia, South Africa, South Korea, Spain, Sri Lanka, Suriname, Swaziland, Sweden, Switzerland, Syria, Taiwan R.O.C., Tanzania, Thailand, Tobago, Trinidad and Tobago, Tunesia, Turkey, Turks and Caicos Islands (B.W.I.), US Virgin Islands, Uganda, Ukraine, United Arab Emirates, United Kingdom, United States, Uruguay, Vanuatu, Venezuela, Vietnam, Yugoslavia (Serbia and Montenegro), Zambia, Zimbabwe.

Careers

On their company website, Aon suggests that "whether you're an experienced professional or just embarking on your career, Aon has an opportunity for you. As the world's premier insurance brokerage and consulting firm, with offices in over 120 countries, Aon offers a tremendous variety of career paths and cultures. Our mind set is truly global. Aon thrives by sharing information across borders while encouraging creativity and independent thinking. We welcome diversity, and we value continuous learning from formal programs and from each other."

You should visit the country-specific websites for information about employment opportunities since all hiring is done locally.

AXA GROUP

Headquarters
   25, avenue Matignon
   75008 Paris
   Phone: (33) (1) 4075 5700
Internet: www.axa.com

The company is made up of business units operating worldwide. Over 50 million individuals and businesses have placed their trust in AXA for home insurance, health insurances, employee benefits, and asset management.

Locations

AXA has offices Algeria, Argentina, Australia, Austria, Belgium, Brazil, Cameroon, Canada, Chile, China Colombia, Czech Republic, France, Germany,

Greece, Guinea, Hong Kong, Hungary, Indonesia, Ireland, Italy, Ivory Coast, Japan, Lebanon, Libya, Luxembourg, Malaysia, Mexico, Monaco, Morocco, New Zealand, Nigeria, Philippines, Poland, Portugal, Portugal, Russia, Saudi Arabia, Senegal, Singapore, Spain, Sweden, Switzerland, Taiwan, Thailand, the Netherlands, Tunisia, Turkey, United Arab Emirates, United Kingdom, United States, Uruguay, Venezuela. The company employs over 130,000 people worldwide.

Careers

The key components of the AXA human resource philosophy are "anticipating the transformations in the organizations and preparing for change, ensuring that everyone has the resources needed to develop skills, making training a top priority, building an organization that is conducive to teamwork, promoting dialogue with managers to understand how to improve performance, drawing on the strength of cultural diversity." The individual group companies are responsible for implementing this policy in their areas of jurisdiction. The company states that its employees "have a clear vision of professional development opportunities within their work unit, their company and the Group. They view mobility as a vital opportunity to gain experience and build expertise."

DION DURRELL

Headquarters
Suite 306, 20 Queen Street West
Toronto, Ontario M5H 3R3
Phone: (416) 408-2626
Internet: www.dion-durrell.com

Dion Durrell is an actuarial and insurance consulting firm. The company creates and implements "innovative strategies that encompass risk financing, insurance management and insurance distribution solutions."

Locations

London (UK), Montreal, Oakbrook Terrace (Illinois), St. Michael (Barbados), and Toronto (HQ).

Careers

Dion Durrell has significant experience in actuarial consulting for the financial sector. The company focuses on strategy rather than on valuation and compliance issues. Their fields of activity include bank insurance, creditor insurance, life actuarial and life insurance consulting, casualty actuarial and general insurance consulting. The company "has a reputation for thinking outside the box and unscrambling complex issues, whether you are seeking effective new ways to save money, to generate revenue or to free up capital in insurance-related matters. Intimacy with current changes in the industry adds value to our strategic advice."

"Working with a blue-chip client list, in areas such as mergers and acquisitions, we have also conducted due diligence assignments, including determinations of fair value and valuation of liabilities, and performed as appointed actuaries and expert witnesses."

ENTEGRIA (UK)
(Entegria Ltd is a Hogg Robinson Company)

Headquarters
42-62 Greyfriars Road
Reading, Berkshire, RG1 1NN
Phone: (44) (0) (118) 958 3683
Internet: www.entegria.co.uk

Entegria is a pensions and employee benefits consultant. The company also provides healthcare consultancy and administration and human resource consultancy services through its specialist arms Remedi and Skillbase.

Locations

Entegria has regional offices in Birmingham, Leeds, London, Reading (HQ), and Waterlooville.

Careers

"Our graduates work alongside both qualified actuaries and other actuarial students at our offices in Reading, London and Leeds. Exposed to a variety of real client issues from the start you will become a valuable member of a motivated, supportive and sociable team. The Company is small enough to provide an innovative and supportive atmosphere to its employees, whilst at the same time offering the opportunities and benefits associated with being part of an international organization—Hogg Robinson."

ERNST & YOUNG

Headquarters
5 Times Square
New York, New York, 10036-6530
Phone: (212) 773-3000
Internet: www.ey.com

According to their website, Ernst & Young offers "a broad array of solutions in audit, tax, corporate finance, transactions, online security, enterprise risk management, the valuation of intangibles, and other critical business-performance issues."

Locations

The company has offices in more than 130 countries and employs about 110,000 people in 670 locations.

Careers

Ernst & Young cites four main reasons why its employees excel when they join the company: its teams, its commitment to learning, its recognition of the importance of work/life balance, and its leadership. The company has a unique culture known as "people first." It is based on the belief that a company can't be great without great people. As a result, Ernst & Young is frequently cited in publications such as *Fortune 500* and *Fortune Magazine* as one of the best places to work.

HEWITT ASSOCIATES

Headquarters
    100 Half Day Road, Lincolnshire, Illinois 60069
    Phone: (847) 295-5000
Internet: www.hewitt.com

Hewitt Associates is a global consulting and outsourcing firm delivering a complete range of human capital management services to companies, including HR and Benefits Outsourcing, HR Strategy and Technology, Health Care, Organizational Change, Retirement and Financial Management, and Talent and Reward Strategies.

Locations

Hewitt Associates have offices around the world, in the Asia-Pacific region, Canada, Europe, Latin America, and the United States. The company delivers services through 86 offices in 37 countries worldwide.

Careers

According to the Hewitt website, chances are that "at Hewitt you'll find a career opportunity that matches your interests, your background, and your goals. Whether you want to work with customers or with technology—behind the scenes or face-to-face—we can give you the opportunity to build the career you always wanted. So think big. The career you're looking for is right here.

"Actuarial science is the single largest professional specialty at Hewitt, with over 400 actuaries in many locations providing services to clients." One of the company's unique aspects is its flat structure. Hewitt believes that success is a shared experience. The company fosters "an environment of growth and learning," and offers flexible career paths.

HYMAN ROBERTSON

Headquarters
Finsbury Tower
103–105 Bunhill Row, London EC1Y 8LZ
Phone: (44) (0) (20) 7847 6000
Internet: www.hymans.co.uk

Hymans Robertson is one of the longest established independent firms of consultants and actuaries in the United Kingdom. The company provides advisory and management services to sponsors and trustees of pension schemes and advice to employers on all aspects of employee benefits.

Locations

The company has offices in Birmingham, Glasgow, and London (HQ).

Careers

Hymans Robertson states that it is a modern and progressive organization with success as one of its core values. According to its website, "professional excellence, commitment and flexibility are key factors in our search for the right candidates and together with generous rewards we can offer the chance to work in a stimulating and challenging environment." One of the company's main fields of expertise is actuarial consultancy.

MELLON FINANCIAL CORPORATION
Human Resources and Investor Solutions (HRIS)

Headquarters
One Mellon Center
Pittsburgh, PA 15258
(412) 234-5000
Internet: www.mellon.com

With more than 3,000 client relationships, the HRIS division of Mellon helps employers provide for the health, welfare and security of an estimated 15 million men and women worldwide. The company's clients include local companies and global corporations, not-for-profit and educational institutions, and numerous state and local governments.

Locations

In addition to its multiple offices in the United States, the HRIS division of Mellon has international offices in Adelaide, Barcelona, Brisbane, Bristol, Brussels, Dublin, Edinburgh, Gouda, Hong Kong, Houston, Ipswich, London, Madrid, Manchester, Melbourne, Mexico City, Montreal, Ottawa, Paris, Perth, Reading, Singapore, Sydney, Toronto, Vienna, Warsaw, and Wiesbaden.

Careers

The business of the HRIS division of Mellon is built around selling our ideas. That's why it's important for them to recruit and retain the top consulting talent in the industry. How do they do that? They create an environment where top consultants want to work. The company knows that its employees are its strongest asset. They therefore provide an excellent work environment, competitive pay and a generous benefits program. One of their slogans is "teamwork with minimum bureaucracy."

MERCER
(A Marsh & McLennan Company)

Headquarters
   Marsh & McLennan
   1166 Avenue of the Americas
   New York, NY 10036-2774
   Phone: (212) 345-5000
Internet: www.mmc.com

Mercer is the consulting business of Marsh & McLennan. It is the world's largest human resources consulting firm. The Marsh & McLennan subsidiaries are NERA Economic Consulting, MERCER Government Human Resource Consulting, MERCER Human Resource Consulting, LIPPINCOTT MERCER Identity and Brand Strategy Consulting, MERCER Investment Consulting, MERCER DELTA Organizational Consulting.

Locations

In addition to its 43 American locations in Albuquerque, Atlanta, Baltimore, Birmingham, Boston, Charlotte, Chicago, Cincinnati, Cleveland, Columbus, Dallas, Deerfield IL, Denver, Detroit, Houston, Indianapolis, Kansas City, Los Angeles, Louisville, Memphis, Milwaukee, Minneapolis, New York, Norwalk, Orange, Philadelphia, Phoenix, Pittsburgh, Portland, Princeton, Richmond, Rochester, Salt Lake City, San Francisco, San Jose, Seattle, St. Louis, Tampa, and Washington DC, Mercer has international offices in over 40 countries around the world: Argentina, Australia, Austria, Belgium, Brazil, Canada, Chile, China, Colombia, Czech Republic, Denmark, Finland, France, Germany, Hong Kong, Hungary, India, Indonesia, Ireland, Italy, Japan, Malaysia, Mexico, New Zealand, Norway, Philippines, Poland, Portugal, Singapore, South Korea, Spain, Sweden, Switzerland, Taiwan, Thailand, the Netherlands, Turkey, the United Kingdom, and Venezuela. The company has more than 15,000 employees.

Careers

"Mercer's business is built on its people: their ideas, their energy, their innovation, their commitment. To attract the best professionals, Mercer strives to be the employer of choice by offering a productive work environment that fosters open communication, trust, mutual respect, teamwork and professional development." The Mercer work experience is distinguished by it shared values, commitment to success through partnership, and flexible career paths and work arrangements. As Mercer puts it: "If you start your career at Mercer, you can expect to learn a lot very quickly through exposure to top professionals, assignment to significant projects, access to tremendous global resources, and the support of great managers and colleagues. No firm in our business offers a wider range of opportunities and services in more locations. Our assignments are challenging. We have high expectations, but that's what makes working here great!" The qualities Mercer expects its employees to have are "a track record of success in university or business, well-developed interpersonal skills, and proven abilities to be effective team players and to handle concurrent demands."

NORMANDIN BEAUDRY

Headquarters
  1130, Sherbrooke Street West, Suite 1100
  Montreal (Quebec) H3A 2M8
  Phone: (514) 285-1122
Internet: www.normandin-beaudry.ca

Normandin Beaudry aims to be the benchmark of actuarial consulting firms for Quebec's enterprises. It is active in pension and savings plans, group benefits, property and casualty insurance and risk management, asset management consulting, and financial commitment valuation.

Location
Montreal

Careers

Normandin Beaudry builds on its distinctive strengths: imaginative proposals, clear communications and profitable solutions for all. Normandin Beaudry draws its strength from six basic principles: "tailor-made teams, client-oriented approach, guarantee of clarity, innovative vision, high-caliber research as well as fair and reasonable fees." The company favors the "early involvement of junior actuaries in various complex projects, encourages them to develop early client relations, counts on experienced professionals who are eager to share their knowledge with new recruits, thus favoring development of skills, its junior actuaries an opportunity to receive guidance from 'mentors' while they are progressing through their enriching and motivating career path, encourages them to become professionally qualified and offers them the required support to reach that goal," and "is constantly looking for bright young professionals and recognizes the full value of a fresh new look at different problems."

PRICE WATERHOUSE COOPERS

Headquarters
   1177 Avenue of the Americas
   New York, New York 10036
   Phone: (646) 471-4000
Internet:www.pwcglobal.com

The worldwide services of PriceWaterhouseCoopers are organized into five main categories: (1) Audit, Assurance and Business Advisory Services, (2) Business Process Outsourcing, (3) Corporate Finance and Recovery Services, (4) Human Resource Services, and (5) Global Tax Services. PriceWaterhouseCoopers employs over 125,000 people in more than 142 countries, "channeling knowledge and value through five lines of service and 22 industry-specialized practices."

Locations

In the United States alone, PriceWaterhouseCoopers has offices in Albany, Atlanta, Austin, Baltimore, Battle Creek, Birmingham, Bloomfield Hills, Boston,

Buffalo, Cambridge, Century City, Charlotte, Chicago, Cincinnati, Cleveland, Columbus, Dallas, Dayton, Denver, Detroit, Florham Park, Fort Lauderdale, Fort Worth, Grand Rapids, Greensboro, Harrisburg, Hartford, Honolulu, Houston, Indianapolis, Irvine (Orange County), Jacksonville, Jersey City, Kansas City, Knoxville, Las Vegas, Lexington, Little Rock, Los Angeles, Louisville, McLean (Tysons Corner), Melville, Memphis, Menlo Park, Miami, Milwaukee, Minneapolis, Montgomery, Montpelier, New Haven, New Orleans, New York (HQ), Ogden, Orlando, Peoria, Philadelphia, Phoenix, Pittsburgh, Portland (Maine), Portland (Oregon), Raleigh, Richmond, Ridgewood, Rochester, Sacramento, Salt Lake City, San Diego, San Francisco, San Jose, Sarasota, Seattle, Spartanburg, St. Louis, Stamford, Syracuse, Tampa, Toledo, Tulsa, Washington, D.C. and West Palm Beach.

Actuaries at PriceWaterhouseCoopers work mainly in the Actuarial and Insurance Management Solutions group. They provide private and public organizations throughout the world with business insurance solutions.

Typical actuarial activities in casualty actuarial consulting involve a "broad range of risk analysis services related to personal and commercial lines, such as automobile liability, general liability and workers' compensation. Services include loss reserving, ratemaking, financial performance and strategy consulting, merger and acquisition valuations, reinsurance program review, and expert witness testimony." The Casualty Actuarial Consulting group of PriceWaterhouseCoopers is the third largest casualty actuarial consulting organization, and the largest casualty actuarial consulting practice of any accounting firm in the United States."

Actuarial life insurance activities at PriceWaterhouseCoopers involve assisting clients "with critical strategic and/or financial planning issues, as well as operational or regulatory compliance aspects of life insurance companies. Services include financial analysis, taxation, litigation support, attestation, product development and many others."

Actuaries at PriceWaterhouseCoopers are also active in insurance operations practice. They deliver a "broad range of claims and underwriting-related services to address insurance-related issues faced by a diverse clientele, including insurers, reinsurers, self-insureds, regulators, captives, capital markets and law firms. Services include claim and underwriting practices reviews, internal controls assessments, regulatory compliance diagnostics, market conduct examinations, litigation and due diligence support, claims portfolio valuations, and Managing General Agency controls reviews."

At the entry level, actuarial candidates are expected to have a Bachelor's or Master's degree and "a strong academic background in actuarial science, applied statistics, financial analysis, insurance, mathematics or related quantitative disciplines." Candidates are also expected to have strong verbal and written communication skills, and to have software skills that include Microsoft Excel, Word and

Access. Moreover, the company expects candidates to be committed to obtaining a Fellowship in the Casualty Actuarial Society. Preferred candidates should at least have passed one of the SOA or CAS examinations.

TILLINGHAST–TOWERS PERRIN
(A Division of Towers Perrin)
Headquarters
335 Madison Avenue
New York, NY 10017-4605
Phone: (212) 309-3400
Internet: www.tillinghast.com/tillinghast

Tillinghast provides actuarial and management consulting to financial services companies and advises other organizations on their self-insurance programs. The company employs over 250 actuaries and several hundred other professionals and is premier actuarial advisor to the insurance industry. Tillinghast operates globally as a single firm with consistent professional standards through a network of 42 offices in 20 countries.

Locations

In the Americas, Tillinghast has locations in Arlington, Atlanta, Bermuda, Boston, Buenos Aires, Chicago, Dallas, Denver, Detroit, Hartford, Jacksonville, Mexico City, Minneapolis, Montreal, New York, Parsippany, Philadelphia, Rio de Janeiro, San Diego, San Francisco, São Paulo, St. Louis, Stamford, Toronto, and Washington, D.C. In Europe and Africa, Tillinghast has offices in Amsterdam, Cape Town, Cologne, Geneva, London, Madrid, Milan, Paris, Rome, Stockholm, and Zürich. Moreover, in Asia, Tillinghast is represented in Hong Kong, Kuala Lumpur, Melbourne, Seoul, Singapore, Sydney and Tokyo.

Careers

The role of new hires depend on the nature of your assignments, their level of experience, and the business practice in which they work. "Employees with undergraduate degrees typically begin in supporting roles on project teams, and take on increased project and client relationship management responsibilities over time. Experienced employees typically begin as project and/or client relationship managers."

TOWERS PERRIN

Headquarters
335 Madison Avenue
New York, NY 10017-4605
Phone: (212) 309-3400
Internet: www.towers.com

Towers Perrin is one of the world's largest global management consulting firms, assisting organizations in managing people, performance and risk.

The firm has provided innovative advice and assistance to large organizations in both the private and public sectors for more than 60 years. The firm's clients include three-quarters of the world's 500 largest companies and three-quarters of the Fortune 1000 largest U.S. companies. Towers Perrin has over 9,000 employees and 78 offices in 23 countries.

Locations

The Towers Perrin offices around the world are grouped into six regions: Africa, Asia/Pacific, Canada, Europe, Latin America and the Caribbean, the United States and Bermuda. In the United States, Towers Perrin has offices in 20 states: Arizona, California, Colorado, Connecticut, Florida, Georgia, Illinois, Massachusetts, Michigan, Minnesota, Missouri, North Carolina, New Jersey, New York, Ohio, Pennsylvania, Texas, Virginia, Washington, and Wisconsin. The company is also represented in Bermuda. The Canadian offices of Towers Perrin are located in Calgary, Mississauga, Montreal, Toronto, and Vancouver. In Africa, Towers Perrin has an office in Johannesburg. In Europe, Towers Perrin has offices in Belgium, France, Germany, Italy, The Netherlands, Spain, Sweden, Switzerland, and the United Kingdom. In addition, the company has offices in China, Japan, Malaysia, Singapore, South Korea, and Australia. In Latin America, Towers Perrin offices are located in Buenos Aires, Rio de Janeiro, São Paulo, and Mexico City.

Careers

The Towers Perrin career section describes job opportunities, the hiring process, campus recruiting events, what students can expect in an interview, actuarial opportunities and how to evaluate job offers. The company website has career profiles of several of its employees describing what it is like to work at Towers Perrin. You must work well on teams with people having diverse perspectives, be willing to be continually challenged. Work opportunities include consulting, corporate, information technology and human resources administration and outsourcing.

WATSON WYATT

Headquarters (UK)
Watson House, London Road
Reigate, Surrey RH2 9PQ, England
Phone: (44) (0) 1737 241144
Headquarters (US)
1717 H Street, NW
Washington, D.C. 20006
Phone: (202) 715-7000
Internet: www.watsonwyatt.com

Watson Wyatt Worldwide is a global consulting firm focused on human capital and financial management. The company specializes in four areas: employee benefits, human capital strategies, technology solutions, and insurance and financial services. Watson Wyatt has more than 6,300 associates in 89 offices in 30 countries.

Locations

In Asia/Pacific, Watson Wyatt has offices in Australia (Melbourne, Sydney), China (Beijing, Hong Kong, Shanghai, Shenzhen), India (Delhi, Kolkata, Mumbai), Indonesia (Jakarta), Japan (Tokyo), Korea (Seoul), Malaysia (Kuala Lumpur), New Zealand (Auckland, Wellington), Philippines (Manila), Singapore, Taiwan (Taipei), and Thailand (Bangkok). Its Canadian offices are in Calgary, Kitchener-Waterloo, Montréal, Ottawa, Toronto, and Vancouver. Watson Wyatt has offices in Europe in Belgium (Brussels), France (Paris), Germany (Düsseldorf, Munich), Hungary (Budapest), Ireland (Dublin), Italy (Milan, Rome), Portugal (Lisbon), Spain (Madrid), Sweden (Stockholm), Switzerland (Zurich), the Netherlands (Amsterdam, Eindhoven, Nieuwegein, Rotterdam) and the United Kingdom (Birmingham, Bristol, Edinburgh, Leeds, London, Manchester, Redhill, Reigate, Welwyn). The Latin American offices of Watson Wyatt are in Argentina (Buenos Aires), Brazil (São Paulo), Colombia (Bogotá), Mexico (Mexico City), and Puerto Rico (San Juan). In the United States, Watson Wyatt's offices are in Atlanta, Boston, Charlotte, Chicago, Cleveland, Columbus, Dallas, Denver, Detroit, Grand Rapids, Honolulu, Houston, Irvine, Los Angeles, Memphis, Miami, Minneapolis, New York, Philadelphia, Phoenix, Portland, Richmond, Rochelle Park, San Diego, San Francisco, Santa Clara, Seattle, St. Louis, Stamford, and Washington, D.C.

Careers

According to *Consulting Magazine*, Watson Wyatt is ranked as one of the 10 best consulting firms to work for. As mentioned on the Watson Wyatt website,

"*Consulting Magazine* recognized Watson Wyatt for its stellar reputation, thought leadership and deep research. It also praised Watson Wyatt's informal family-oriented culture that rewards creativity and hard work."

INSURANCE COMPANIES

In this appendix, we profile some leading insurance companies. All of these companies employ actuaries. The list is incomplete and somewhat random since there are simply too many companies to discuss in this guide. The directory of insurance companies on the Internet alone, found at *www.iiin.com*, lists close to 3,000 companies, grouped into health, life, property and casualty, reinsurance, specialty insurance, and title insurance, with over 300 of these companies represented internationally. In addition, most if not all of the large international banking institutions now have actuarial divisions. Only one or two of them are included in the section since the employment options that they provide tend to be quite similar.

The descriptions that follow are merely meant to be starting points for your career search. As you browse through them, you should get a sense of what it is like to work as an actuary for different kinds of insurance companies. For this reason, the individual profiles stress different aspects of employment: global mobility, daily tasks, required qualifications, company philosophy, working conditions, and so on. The quoted material can be found on the websites of the respective companies. You should consult these sites for more complete information. Job descriptions included with some company profiles do not indicate that these jobs are still available. They are meant to illustrate different aspects of actuarial life, and to give real-world examples of the main theme of this guide, which stresses the bond between mathematics, business, and statistics upon which actuarial careers are built.

AETNA

Headquarters

Aetna Inc.
151 Farmington Avenue
Hartford, Connecticut 06156
Phone: (860) 273-0123
Internet: www.aetna.com

Aetna is one of the leading providers of health, dental, group life, disability, and long-term care benefits in the United States.

Locations

Hartford (Connecticut), with field offices throughout the United States.

Careers

Aetna actuaries are the financial architects of the company. While most of Aetna's career opportunities start with the company's actuarial training program, they occasionally hire experienced, professional actuaries. "If you are on your way to achieving your goal—Fellow of the Society of Actuaries—or already have the professional designation of Fellow, we'd like to hear from you."

"Aetna is dedicated to helping people manage what matters most in their lives—their health and well-being. As a leading provider of employee benefits, we are proud of the range of benefits we extend to our own employees. We view our employee benefits as more than mere coverage—it's a way to say 'thank you' to our employees for choosing to give us their time, passion and hard, earnest work."

AIG

Headquarters

70 Pine Street
New York, New York 10270
Phone: (212) 770-7000
Internet: www.aig.com

AIG is a leading international auto, health, life insurance and financial services organization based in the United States.

Locations

AIG is one of the world's leading international insurance and financial services organization. It operates in approximately 130 countries.

Careers

"Whether your experience is in accounting and finance, underwriting, actuarial or technology, if you're a problem solver, facilitator and an out-of-the-box thinker, the AIG companies may be where you belong."

ALLIANZ GROUP

Headquarters
    Königinstrasse 28
    Munich 80802
    Phone: (49) (0) (89) 3800-0
Internet: www.allianzgroup.com

Allianz Group is a multinational group of 700 companies and more than 181,000 employees worldwide. Using decentralized management, Allianz Group possesses high levels of competency in local markets. This allows for maximum adaptability and a strengthening of its local resources. In a changing environment, the companies of Allianz Group combine stability and continuity with the ability to act strategically.

Locations

The US members of the Allianz Group include Allianz Dresdner Asset Management (Newport Beach, California), Allianz Hedge Fund Partners (San Francisco, California), Allianz Insurance Company (Burbank, California), Allianz Life Insurance Company of North America (Minneapolis, Minnesota), Allianz Risk Transfer, Inc. (New York, New York), Cadence Capital Management (Boston, Massachusetts), Dresdner RCM Global Investors (San Francisco, California), Fireman's Fund AgriBusiness (Overland Park, Kansas), Fireman's Fund Insurance Company (Novato California), Fireman's Fund McGee Underwriters (New York, New York), Interstate Insurance Group (Chicago, Illinois), NFJ Investment Group (Dallas, Texas), Nicholas-Applegate Capital Management (San Diego, California), Oppenheimer Capital (New York, New York), PIMCO (Newport Beach, California), PIMCO Advisors Distributors (Stamford, Connecticut), and PIMCO Equity Advisors LLC (New York, New York).

Careers

"There are 181,000 outstanding reasons for our success. And you can be another. We have been growing with our staff—and because of them—for more than 100 years. Allianz Group is a place where talent is given the chance to flourish."

ALLIANZ INSURANCE COMPANY

Headquarters
> 2350 Empire Avenue
> Burbank, California 91504-3350
> Phone: (818) 260-7500

Internet: www.aic-allianz.com

Allianz is one of the world's leading international insurance companies. Through the Allianz Group network, AIC has a global reach in 77 countries. In the United States, the company's activities are mostly in the areas of property and casualty insurance for business and individuals and in asset management. AIC is a leading carrier for large corporations and their global risks.

Locations

In the United States, AIC has regional offices in New York, Chicago, Houston and Atlanta.

Careers

An entry-level technical assistant with previous experience in RMS [risk management solutions] and CAT [catastrophe coverage] modeling is expected to have the qualifications that include "strong communication, multi-tasking, and organizational skills, intermediate to advanced MS Excel skills, knowledge of property insurance concepts, mathematical aptitude, strong analytical skills, ability to perform within time constraints, and strong written, verbal and telephone communication skills."

ALLIANZ LIFE INSURANCE COMPANY

Headquarters
> P.O. Box 1344
> Minneapolis, Minnesota 55416-1297
> Phone: (800) 950-5872

Internet: www.allianzlife.com

Allianz Life offers fixed and variable annuities, universal life insurance, and long-term care insurance. It is among the top insurance providers in North America.

Location

Minneapolis (Minnesota)

Careers

Here are two job descriptions for different levels of actuaries employed by Allianz Life:

An actuary "will provide accurate evaluation and communication of the financial implications of future contingent events to facilitate the appropriate management of returns and risks in the business unit. Corporate actuarial consists of product development, risk management and financial reporting." The actuary "is responsible for research and development of valuation and financial reporting requirements for life and annuity products. This position will also assist product development and other financial actuaries to ensure optimal valuation approaches are implemented for new products and valuation pronouncements." The actuary is expected to be an FSA and MAAA with actuarial experience in the life insurance industry and deterministic and stochastic modeling, and have "strong knowledge of life and annuity insurance products and valuation requirements under statutory, GAAP [generally accepted accounting principles] and tax methodologies, have the ability to work independently and design solutions to a variety of financial problems, strong verbal and written communication skills with persons at all levels of experience and expertise," and have the "ability to apply actuarial valuation principles within regulatory frameworks, financial reporting standards and risk profiles to meet company needs."

An associate actuary, on the other hand, is expected to have a Bachelor's degree in actuarial science, mathematics, or related field, two or more SOA examinations passed, with two or more years of actuarial work experience, with an investments and ALM [asset and liability management] background, have analytical thinking skills, proficiency with computer software including TAS and Excel, effective verbal and written communication skills, a general understanding of actuarial methods, tools and issues of business area, the ability to manage smaller project, have good organizational and time management skills, the ability to verify and document work, to work independently and in teams.

ALLSTATE

Headquarters
   Allstate Insurance Company
   2775 Sanders Rd. Ste F7
   Northbrook, Illinois 60062
     Phone: (800) 427-9389
Internet: www.allstate.com

Allstate is a major auto, home, life, and business insurance company in the United States and Canada.

Locations

Allstate Insurance Company, Allstate Indemnity Company, Allstate Life Insurance Company, Allstate Property and Casualty Insurance Company, Glenbrook Life and Annuity Company, Northbrook Life and Annuity Company, all headquartered in Northbrook, Illinois. Allstate New Jersey Insurance Company (Bridgewater, New Jersey). Allstate Life Insurance Company of New York (Hauppauge, New York), Allstate Floridian Insurance Company and Allstate Floridian Indemnity Company (St. Petersburg, Florida), Allstate County Mutual Insurance Company and Allstate Texas Lloyd's (Irving, Texas), American Heritage Life Insurance Company (Jacksonville, Florida), Lincoln Benefit Life Company (Lincoln, Nebraska).

Careers

"At Allstate, actuaries play a vital role in developing the property and casualty products that our agents sell to customers. An actuary's role includes everything from researching new product concepts and product enhancements to recommending and implementing pricing changes in each state. Allstate actuaries use their unique combination of problem solving, analytic, and communication skills to forecast the costs, expenses and income associated with providing insurance coverage."

AMERICAN RE
(A Member of the Munich Re Group)

Headquarters
    555 College Road East
    Princeton, NJ 08543
    Phone: (609) 243-4200
Internet: www.amre.com

American Re specializes in business reinsurance with a focus on small and midsize companies.

Locations

Princeton (New Jersey) and branch offices throughout the United States.

Careers

"We are a recognized leader in the industry because our philosophy places primary value on relationships, which also extends to our most important asset, our employees. American Re has a corporate culture that supports our employees professionally and promotes teamwork, emphasizes communication, and values the contributions of all. We endeavor to provide our staff with career development, skill enhancement, personal reward and satisfaction. We offer a comprehensive benefit program and a stimulating, challenging, and employee-friendly work environment." At the level of Vice-President, an employee of American Re is expected to be a Fellow of CAS, SOA, and a member of the American Academy of Actuaries (MAAA). The employee is also expected to have "strong technical actuarial skills, strong software skills, including the ability to manage programmers, strong interpersonal skills, good time management skills, excellent oral and written communication skills."

AVIVA

Headquarters
St Helen's, 1 Undershaft
London, EC3P 3DQ
Phone: (44) (0) (20) 7662 7122
Internet: www.aviva.com

Aviva is the world's seventh-largest insurance group and the biggest in the UK. The company was created by merger of CGU and Norwich Union in 2000. Aviva is one of the leading providers of life and pensions insurance in Europe and has substantial businesses elsewhere around the world. Its main activities are long-term savings, fund management and general insurance.

Locations

In the United Kingdom, Aviva has offices in London (HQ), Norwich, and York. Its international offices are in Australia, Belgium, Brunei, Canada, China, Cyprus, Czech Republic, France, Germany, Gibraltar, Greece, Hong Kong, Hungary, India, Indonesia, Ireland, Italy, Japan, Lithuania, Luxembourg, Malaysia, Malta, Netherlands, Philippines, Poland, Singapore, Spain, Thailand, Turkey, and the United States.

Careers

Aviva maintains a systematic global employee training and development program, based on the philosophy that "effective training and development helps the company to attract and retain high quality people, support them in reaching their potential and building the capabilities necessary to succeed in a changing and challenging environment."

AVIVA CANADA
(A Subsidiary of Aviva plc (UK))

Headquarters
    Aviva Canada Inc.
    2206 Eglinton Avenue East
    Scarborough, Ontario, M1L 4S8
    Phone: (416) 288-1800
Internet: www.cgu.ca

Aviva Canada is one of the largest property and casualty insurers in Canada.

Locations

Calgary, Dartmouth, Drummondville, Edmonton, Hamilton, London, Montreal, Ottawa, Québec, St. John, Toronto (HQ), Vancouver, and Winnipeg.

Careers

The strength of Aviva is built on its corporate values: "integrity, commitment, excellence in execution, teamwork and performance and results oriented." Anyone interested in a career at Aviva benefits from and must buy into these values.

BLUE CROSS BLUE SHIELD
(An Association of Independent Blue Cross Blue Shield Companies)

Headquarters
    225 North Michigan Avenue
    Chicago, Illinois 60601-7680
    Phone: (312) 297-6000
Internet: www.bcbs.com

Blue Cross Blue Shield is the oldest and largest health insurance organization in America.

Locations

Alabama, Alaska, Arizona, Arkansas, California, Colorado, Connecticut, Florida, Georgia, Hawaii, Idaho, Illinois, Indiana, Iowa, Kansas, Kentucky, Louisiana, Maine, Massachusetts, Michigan, Minnesota, Mississippi, Missouri, Montana, Nebraska, Nevada, New Hampshire, New Jersey, New Mexico, North Carolina,

North Dakota, Ohio, Oklahoma, Oregon, Pennsylvania, Puerto Rico, Rhode Island, South Carolina, South Dakota, Tennessee, Texas, Utah, Vermont, Virginia, Washington, West Virginia, Wisconsin, Wyoming, and Canada.

Careers

An actuarial analyst working for one of the member companies is expected to be able to "provide support for insurance pricing, provider contracting, financial reporting, and reserving. Develop and maintain computer programs, prepare rate and provider reimbursement studies and other statistical analyses, produce various reports, and perform other general actuarial functions. Provide technical support and develop work plans for projects to be completed by self with possible support of others. Has obtained broad understanding of the general objectives of the actuarial department and basic understanding of insurance risks." The minimum qualifications asked for are: "A college degree in mathematics, statistics, or computer programming or equivalent experience. Strong analytical, problem solving, and troubleshooting skills. Effective oral and written communication skills. Ability to work independently and as a member of a team. Attention to detail for checking own work as well as others' work. Ability to take direction well."

CANADA LIFE

Headquarters
> The Canada Life Assurance Company
> 330 University Avenue
> Toronto, Ontario M5G 1R8 Canada
> Phone: (416) 597-1440

Internet: www.canadalife.ca

Canada Life is one of the largest insurance companies in Canada. It is one of Canada's top life insurers. Canada Life provides services to more than ten million policyholders throughout Canada, the United States, the United Kingdom and Ireland.

Locations

Canada Life is located in Canada, the United States, the Bahamas, Brazil, Germany, the Isle of Man, the Republic of Ireland, Puerto Rico and the United Kingdom.

Careers

An actuary working for Canada Life in the United States, for example, would have to be self-motivated, have strong technical, analytical, organizational and communication skills and, at the more senior level, be a Fellow of the SOA.

CIGNA

Headquarters
One Liberty Place, 1650 Market Street
Philadelphia, Pennsylvania 19192
Phone: (215) 761-1000
Internet: www.cigna.com

CIGNA is an employee benefits company in the United States and selected markets around the world. It provides financial services such as discount brokerage services for investors and retirement planning, disability, life and accident group insurance.

Locations

CIGNA is represented in Brazil, Chile, Indonesia, Japan, Korea, Spain, United Kingdom, the United States, and Taiwan.

Careers

CIGNA seeks to hire top performers. It look for people who are "motivated and results-driven, energized and hard-working." The company's commitment to its employees is formalized in its *employee value proposition*, "a thoughtful, well-defined statement geared toward building successful careers within a successful company."

COMBINED INSURANCE COMPANY (An Aon Subsidiary)

Headquarters
1000 N. Milwaukee Ave.
Glenview, IL 60025
Telephone: (847) 953-2025
Internet: www.combined.com

Combined Insurance Company of America is a subsidiary of Aon Corporation. It is the largest of Aon's insurance underwriting companies and services five million policyholders worldwide through a sales force of over 7,000 people throughout North America, Europe and the Pacific.

Locations

Combined and its subsidiaries operate in the following countries and territories: Australia, Canada, Germany, New Zealand, Portugal, Puerto Rico, Republic of Ireland, United Kingdom, United States, and US Virgin Islands.

Careers

"Since 1919, Combined Insurance Company of America has been bringing qual- ity supplemental accident, disability, health and life insurance to individuals and families across the United States and seven other countries. Combined is the largest consumer insurance underwriting company of Aon Corporation, the world's premier insurance brokerage, consulting services and consumer insur- ance underwriting organization." Combined provides career opportunities in five different areas: accident, life, health, seniors, and worksite solutions.

CONVERIUM
(Formerly Zurich Re)

Headquarters
　Baarerstrasse 8
　6300 Zug, Switzerland
　Phone: (41) (0) (1) 639-9335
Internet: www.converium.com

Converium is a global reinsurer, employing more than 800 people in 22 offices around the world. Its services are provided through Converium Zurich, Con- verium Cologne, Converium North America and Converium Life, and a world- wide network of locally operating units.

Locations

In Europe, Converium has offices in Cologne, Guernsey, London, Milan, Paris, Zug, Zurich. The North American offices of the company are in Atlanta, Bermuda, Chicago, New York, Orange County, San Francisco, and Stamford. In Latin America, Converium has offices in Buenos Aires, Mexico City, and São Paulo. In Asia/Pacific, the company is represented in Kuala Lumpur, Labuan, Singapore, Sydney and Tokyo.

Careers

Through its local offices, Converium is active in the following broad areas of insurance: accident and health, agribusiness, automobile liability, aviation and space, casualty clash, credit and surety, e-commerce, engineering, excess and surplus liability, general third party liability, intellectual property, life marine, professional liability, property and catastrophe, risk strategies, weather risk man- agement, and workers compensation. The company stresses the importance of providing a balance between work and life for its employees and has a variety of programs and benefit structures in place to make this happen.

DESJARDINS GROUP

Headquarters
100 Avenue de Commandeurs
Lévis, Québec G6V 7N5
Phone: (418) 838-7870
Internet: www.desjardins.com

The Desjardins Group has subsidiaries active in various sectors of the financial services industry: Desjardins Financial Security (life and health insurance), Desjardins Group General Insurance (property and casualty insurance), Desjardins Specialized Financial Services Management (design and distribution of mutual funds, and trust services), Desjardins Securities (securities brokerage) and Elantis Investment Management (investment management).

Locations

Montreal, Québec, and, through its affiliations, in other Canadian provinces.

Careers

Desjardins employs specialists in a wide range of fields. In the actuarial field, the company is active in actuarial analysis, research and development of actuarial services, actuarial statistics and ratemaking. Desjardins also employs specialists in risk and credit management, general insurance, health and life insurance, economics, finance and accounting, and capital markets, funds, and investments.

EVEREST REINSURANCE GROUP

Headquarters
477 Martinsville Road
P.O. Box 830
Liberty Corner, New Jersey 07938-0830
Phone: (908) 604-3000
Internet: www.everestregroup.com

Everest Reinsurance is a world leader in property and casualty reinsurance and insurance.

Locations

In addition to its headquarters in New Jersey, Everest has offices in Barbados, Bermuda, Brussels, Chicago, London, Miami, New York, Oakland, Singapore, and Toronto.

Careers

Most job opportunities are in their corporate headquarters in New Jersey.

FARMERS INSURANCE GROUP

Headquarters
4680 Wilshire Blvd.
Los Angeles, California 90010
Phone: (208) 239-8400
Internet: www.farmers.com

Farmers Insurance Group of Companies is the third-largest writer of both private passenger automobile and homeowners insurance in the United States. The company operates in 41 states and has approximately 18,000 employees.

Locations

Los Angeles (California) and field offices throughout the United States.

Careers

"If you are a professional in information technology, accounting, actuarial, claims, marketing, communications, auditing, legal, administration, human resources or underwriting, Farmers has a career opportunity for you."

FRIENDS PROVIDENT INTERNATIONAL

Headquarters
Royal Court, Castletown
Isle of Man, British Isles, IM9 1RA
Phone: (44) (0) (1624) 821212
Internet: www.fpinternational.com

Friends Provident International is one of the oldest offshore life companies in the world. It specializes in delivering high quality offshore investment products and services to the international community.

Location

Isle of Man

Careers

The company offers high school and college internship programs both in generalist fields and in specific area. It provides internships for local schools and colleges as part of tertiary level qualifications. The company also offers "university sponsorship for courses related to technical areas, such as information technology, finance, actuarial and marketing."

GE ERC
(A General Electric Company)

Headquarters
>5200 Metcalf
>P.O. Box 2991
>Overland Park, KS 66201-1391
>Phone: (913) 676-5200

Internet: www.ercgroup.com

ERC is the world's fourth-largest reinsurer, and provides insurances services in property and casualty, life, healthcare, and professional liability insurance and reinsurance, as well as other risk management services.

Locations

The main offices ERC are located in Chicago, Fort Wayne, Hartford, New York, and Overland Park in the United States and London and Munich in Europe. In addition, the company has regional offices in Asia/Pacific in Australia, China, Hong Kong, Japan, Malaysia, New Zealand, and Singapore, in Denmark, France, Greece, Germany, Ireland, Israel, Italy, Lebanon, Luxembourg, Poland, Spain, Switzerland, and the United Kingdom in Europe, Argentina, Brazil, Mexico, Puerto Rico in Latin America, and California, Canada, Colorado, Connecticut, Florida, Georgia, Illinois, Indiana, Kansas, Kentucky, Massachusetts, Michigan, Minnesota, Missouri, New Mexico, New York, North Carolina, Ohio, Pennsylvania, Puerto Rico, Virginia, Texas, Washington, and Wisconsin in North America.

Careers

GE ERC is a large global company providing exciting and innovative employment opportunities around the world. The company recruits at the junior level through its extensive internship program.

GENERAL COLOGNE RE

Headquarters
>Theodor-Heuss-Ring 11
>Sedanstr. 8
>50668 Cologne
>Phone: (49) (0) (221) 9738-0

Internet: www.gcr.com

The General Cologne Re is a leader in global reinsurance and related risk assessment, risk transfer, and risk management operations. The company has over 3,900 employees in 30 countries around the world. The flagship domestic subsidiary, General Reinsurance Corporation, is the largest property/casualty company in North America. The company is part of the Berkshire Hathaway organization.

Locations

The North American offices of the company are in Atlanta, Boston, Charlotte, Chicago, Columbus, Dallas, Hartford, Kansas City, Los Angeles, Montreal, New York, Orlando, Philadelphia, Phoenix, San Francisco, Seattle, St. Paul, Stamford, and Toronto. In Latin America, the company has offices in Buenos Aires, Mexico City, and São Paulo. In Europe, the company is represented in Cologne (HQ), Copenhagen, Dublin, Hamburg, London, Madrid, Manchester, Milan, Moscow, Paris, Riga, Vienna, and Warsaw. The South African offices of the company are located in Cape Town and Johannesburg. In addition, the company has offices in Asia and the Pacific in Auckland, Beijing, Brisbane, Hong Kong, Melbourne, Perth, Seoul, Shanghai, Singapore, Sydney, Taipei, and Tokyo.

Careers

General Cologne Re operates on a global basis and offers a wide range of career opportunities. The company hires university graduates with degrees in actuarial science, economics, mathematics, computer science, accounting, law, engineering, and liberal arts. It places a high value on good academic credentials, relevant employment experience, client/marketing skills, a sense for international business, software skills, language skills, the capacity to work in a team environment, and expects a high degree of energy and creativity.

HANNOVER RE

Headquarters
   Karl-Wiechert-Allee 50
   30625 Hannover, Germany
   Phone: (49) (0) (511) 56 040
Internet: www.hannover-re.com

Hannover Re is a reinsurance company. It provides insurance for insurance companies.

Locations

Hannover Re has offices around the world. In Europe, it has offices in France, Germany (HQ), Ireland, Italy, Spain, Sweden, and the United Kingdom. In North America, its offices are in Bermuda, Canada, Mexico, and the United States (Itasca, New York, Orlando, Los Angeles). In Africa, the company has offices in Mauritius and South Africa, and in Asia/Pacific, it has offices in Australia, China (Hong Kong and Shanghai), Japan, Korea, Malaysia, and Taiwan.

Careers

"The job profiles in our company are just as diverse as the reinsurance business itself. There is no single qualification or degree that makes an applicant *perfect* for us. Generally speaking, successful completion of an insurance training program (apprenticeship) is a good starting point. For university graduates, depending on the area of employment, degrees in business administration, economics, mathematics, law, and even meteorology are of interest to us. What will be crucial to your success in our company is your ability to familiarize yourself with a broad range of topics and react flexibly. Needless to say, we also value qualities indispensable in the modern business environment: creativity, team skills, individual initiative, dynamism, the power of persuasion and determination. There is, however, something which we prize even more highly than these standards: you should be open-minded towards people from a highly diverse range of cultural backgrounds, and—if possible—you should speak one or more foreign languages in addition to possessing a very good command of English."

HARTFORD FINANCIAL SERVICES GROUP

Headquarters
    690 Asylum Avenue
    Hartford, Connecticut 06115
    Phone: (860) 547-5000
Internet: www.thehartford.com

The Hartford has two divisions: Hartford Life and Hartford Property & Casualty. The company offers investment products, individual life insurance, group benefits, and property and casualty insurance.

Locations

Hartford (Connecticut) and regional offices throughout the United States.

Careers

The Hartford uses technology to enhance the quality of its services. It is a pioneer in the web-based delivery of financial services is constantly updating its technology and service standards. "As a leader in insurance, asset management and financial service products, we offer professionals every possibility for growth. And whether we're helping customers or building careers, we're experts at creating the kind of advantages that help people reach their goals."

ING GROUP

Headquarters
    Amstelveenseweg 500
    1081 KL Amsterdam
    Phone: (31) (0) (20) 541 54 11
Internet: www.ing.com

ING is a global financial institution active in banking, insurance and asset management. More than 100,000 people work for ING in 65 countries in virtually every area of the financial services industry.

Locations

The ING Group is represented around the world. In addition to its offices in the Netherlands, ING has offices in Argentina, Aruba, Austria, Belgium, Brazil, Brazilian Virgin Islands, Bulgaria, Canada, Chile, China, Cuba, Czech Republic, France, Germany, Greece, Hong Kong, Hungary, India, Indonesia, Ireland, Italy, Japan, Kazarkhstan, Luxembourg, Macau (China), Malaysia, Mauritius, Mexico, Monaco, Netherland Antilles, New Zealand, Norway, Peru, Philippines, Poland, Portugal, Romania, Russian Federation, Serbia and Montenegro, Singapore, Slovak Republic, South Africa, South Korea, Spain, Switzerland, Taiwan, Thailand, Turkey, Ukraine, United Arab Emirates, United Kingdom, Venezuela, and Vietnam.

Careers

As a global company, ING offers a world of opportunities for people with enthusiasm, talent and ambition.

JOHN HANCOCK

Headquarters
200 Clarendon Street
Boston, Massachusetts 02116-5021
Phone: (617) 572-6000
Internet: www.jhancock.com

John Hancock is one of the largest providers of a full range of insurance and investment products and services of the United States. Its products include annuities, individual and group long-term care insurance, and individual and group life insurance.

Locations

John Hancock operates primarily in the United States, Canada and the Pacific Rim (China, Indonesia, Malaysia, the Philippines, Singapore, and Thailand). Its North American locations are in Albuquerque, Boston (HQ), Halifax, and Los Angeles. The company also has European offices in Brussels, Dublin, and London, and is one of small number of insurance companies licensed to operate in China.

Careers

John Hancock champions ongoing education, "whether it be at our on-site Education Center—which includes a full range of Technical Education courses, industry/technology certification programs, instructional programs on financial service products, and extensive management training—or through external educational institutions the cost of which is considered under our Tuition Award program." The company's professional job opportunities range "from the traditional actuarial, accountant, auditing, claims processing, customer service representatives, finance, law, marketing, money management, pension services, portfolio management, real estate investments, risk management and underwriting positions to the more unique child care providers, community relations, corporate television, corporate education, electronic publishing, graphics, space planning, and voice communications."

LONDON LIFE
(Subsidiary of Great-West Life)

Headquarters
255 Dufferin Avenue
London, Ontario, N6A 4K1
Phone: (519) 432-5281
Internet: www.londonlife.com

Together with Great-West, London Life serves the financial security needs of nine million people across Canada. London Life participates in international markets through London Reinsurance Group, a supplier of reinsurance in the United States and Europe.

Locations

London (Ontario) and regional offices throughout Canada.

Careers

Actuaries are key members of the London Life team. They provide "expert advice on a wide range of business initiatives including the design of financial products, investments, information technology, planning and marketing of products, strategic risk measurement, and almost every other aspect of work in the organization." To meet these needs, London Life offers a development program for actuarial students. "If you are actively pursuing Fellowship in the Canadian Institute of Actuaries (FCIA) designation, our program provides the opportunity to gain work experience in different areas of the company while you are studying for the Society of Actuaries examinations."

MANULIFE FINANCIAL

Headquarters
  500 King Street North
  Waterloo, Ontario N2J 4C6
  Phone: (519) 747-7000
Internet: www.manulife.com

Manulife Financial is a leading Canadian-based financial services company offering annuity, life insurance, pension, and individual wealth management products.

Locations

Manulife Financial operates in 15 countries and territories worldwide. In Canada, Manulife Financial has offices in Kitchener, Montreal, Toronto, and Waterloo. In the United States, the company has offices in Massachusetts. In addition, Manulife Financial conducts business in China, Hong Kong, Indonesia, Japan, Philippines, Singapore, Taiwan, and Vietnam.

Careers

An actuarial analysis at Manulife Financial you must have a Bachelor's degree, strong mathematical and spreadsheet skills, familiarity with basic finance and economic principles. Work may involve working with spreadsheet programs, running of reports from computer applications, and communicating the results through documentation and meetings to appropriate departments.

A product actuary might be responsible for the design, development, profitability analysis and implementation of annuity and retirement income products, monitor the profitability of new products and ensure that these products comply with Federal and State regulations. A product actuary must be a Fellow of the Society of Actuaries and have the ability "to influence others beyond formal authority." Financial understanding of product profitability, excellent oral and written communication skills, excellent project management, problem-solving and analytical skills and leadership ability, people development and motivational skills would also be required.

MARITIME LIFE
(A subsidiary of John Hancock
Financial Services of Boston)

Headquarters
  7 Maritime Place
  Halifax, Nova Scotia E3J 2X5
  Phone: (902) 453-4300
Internet: www.maritimelife.ca

Maritime Life is a Canadian company providing service in life insurance, disability and critical illness insurance, investment, pensions, group benefits, association plans, and institutional investment.

Locations

Calgary, Halifax (HQ), Kitchener, Montreal, Oakville, Toronto, Vancouver

Careers

The actuarial and internship student programs at Maritime Life attract the best and the brightest. Through these programs the company nurtures the careers of potential future employees. Maritime Life provides actuarial students with a work environment where they can apply the knowledge and skills acquire through their studies.

MELOCHE MONNEX
(Member of the TD Bank Financial Group)

Headquarters
  2161 Yonge Street
  Toronto, Ontario M4S3A6
  Phone: (416) 484-1112
Internet: www.melochemonnex.com

Meloche Monnex is the leading organization in affinity marketing in Canada and the second largest direct insurer. The company offers home and automobile insurance to members of professional associations, university organizations, select employer groups, to clients of TD Bank Financial Group, and to some extent, the general public. It provides advice and assistance in home and automobile insurance, travel insurance and insurance for small enterprises.

Locations

Calgary, Edmonton, Halifax, Montreal, Toronto (HQ)

Careers

In addition to being technically competent, actuarial employees at Meloche Monnex must have excellent communication, computer, and language skills.

MUNICH RE

Headquarters
  Koeniginstrasse 107
  Munich 80802
  Phone: (49) (0) (89) 38 91 0
Internet: www.munichre.com

Munich Re is an international reinsurance company with more than 60 offices and subsidiaries worldwide.

Locations

North America: Atlanta, Princeton, Montreal, Toronto, Vancouver. Latin America: Bogotá, Buenos Aires, Caracas, Mexico City, Santiago de Chile, São Paulo. Europe: Athens, Geneva, London, Madrid, Milan, Moscow, Munich (HQ), Paris, Warsaw. Africa, Near and Middle East: Accra, Durban, Harare, Johannesburg, Capetown, Nairobi, Port Louis, Tel Aviv. Asia and Australasia: Auckland,

Brisbane, Hong Kong, Mumbai, Beijing, Perth, Shanghai, Seoul, Singapore, Sydney, Taipei, and Tokyo.

Careers

Mathematicians at Munich Re are considered "today's prophets." They work in life reinsurance, health reinsurance, non-life reinsurance, and IT [information technology].

"It is the business of insurance companies to bear risks by promising to pay financial compensation in the event of a loss. Such financial compensation is given if someone suffers loss or damage covered by an insurance policy they have taken out. Insurance companies, or primary insurers, that assume the risks of the original risk carriers (mostly private individuals or businesses). They themselves thus become risk carriers and therefore require insurance, this form of coverage being known as reinsurance."

Reinsurers must have a widely diversified range of mathematical expertise. They might be called 'today's prophets' because they aim, for example, to determine what the probability of occurrence of various types of loss will be and to predict how a whole package of insurance policies is likely to develop in the future." Account managers at Munich Re serve as interfaces between Munich Re and the insurance companies. "As consultants for all matters related to life insurance, it is their duty to convey to the emerging markets the experience and knowledge that the company's specialists have acquired throughout the world. They deal with pricing, selecting and introducing new products, tax issues as well as the setting of terms and conditions, assessment of risks, distribution of insurance, and so on."

NEW ENGLAND FINANCIAL

Headquarters
  501 Boylston St.
  Boston, Massachusetts 02116-3769
  Phone: (617) 578-2000
Internet: www.nefn.com

Through its national network of professional financial representatives and firms New England Financial offers products that include personal and business financial planning, life and disability insurance, individual and small-group health insurance, executive benefits, tax-qualified retirement plans and employee benefits.

Locations

New England Financial has local marketing firms throughout the United States.

Careers

At its headquarters in Boston and in local marketing firms, New England Financial employs specialists including annuities, disability income, long-term care, retirement planning, and voluntary benefits.

OPTIMUM GENERAL

Headquarters
425 de Maisonneuve Blvd. West
Montreal, Quebec H3A 3G5
Phone: (514) 288-8711
Internet: www.optimum-general.com

Optimum General is a Canadian company that underwrites property and casualty insurance through four subsidiaries: Optimum West, Optimum Frontier, Optimum Insurance, and Optimum Farm. The Company is active in four main insurance lines: automobile, personal property, commercial property and liability insurance. Optimum General and its subsidiaries have approximately 260 employees.

Locations

Optimum Farm: Trois Rivières; Optimum Frontier: Halifax, Moncton, North Bay, Toronto, Winnipeg; Optimum Insurance: Montreal, Quebec; Optimum West: Edmonton, Vancouver

Careers

Optimum General employs both English and French-speaking P/C actuaries, depending on location.

PACIFIC LIFE

Headquarters
700 Newport Center Drive
Newport Beach, California 92660
Phone: (949) 219-3011
Internet: www.pacificlife.com

Pacific Life provides life and health insurance products, individual annuities, mutual funds, group employee benefits, and offers to individuals, businesses, and pension plans a variety of investment products and services.

Locations

Newport Beach and Irvine, California. The company has business relationships with 68 of the 100 largest U.S. companies.

Careers

"If you are looking for new challenges that will take your career further, consider Pacific Life. Whether you are just beginning your career, contemplating a career change, or are a seasoned professional looking for a new opportunity to expand your career, Pacific Life offers a variety of opportunities throughout our company. Our competitive salaries, strong bonus plans, outstanding benefits, excellent training, a business casual dress environment plus many other incentives make up the culture that is Pacific Life." Pacific Life stresses commitment to excellence, employee development, the use of cutting-edge technology, and community involvement.

PRUDENTIAL FINANCIAL

Headquarters
    751 Broad Street
    Newark, New Jersey 07102-3777
    Phone: (973) 802-4291
Internet: www.prudential.com

Prudential Financial companies serve individual and institutional customers worldwide and include The Prudential Insurance Company of America, one of the largest life insurance companies in the United States ("The Rock"). These companies offer a variety of products and services, including life insurance, property and casualty insurance, mutual funds, annuities, pension and retirement related services and administration, asset management, securities brokerage, banking and trust services, real estate brokerage franchises, and relocation services.

Locations

Prudential Financial has a global presence in the insurance industry. The company is represented in Argentina, Belgium, Brazil, Canada, Chile, France, Germany, Hong Kong, Ireland, Japan, Luxembourg, Italy, Japan, Mexico, Monaco, Netherlands, Philippines, Poland, Puerto Rico, Singapore, South Korea, Spain, Switzerland, Taiwan, the United Kingdom, United Arab Emirates, and Uruguay.

Careers

"It's a journey beyond the expected: an experience that can move billions of euros, rupees or yen on any given day lead you across ten thousand acres of pristine forest land, revitalize entire neighborhoods, and carry your ideas to virtually every corner of the globe. It's a career with Prudential. And if you think you know what lies ahead—think again. At Prudential, we recognize that any single opportunity can lead to a thousand different destinations. In both the business we conduct and the careers we build, we are determined to explore them all."

RBC INSURANCE

Headquarters
> 6880 Financial Drive
> West Tower
> Mississauga, Ontario L5A 4M3
> Phone: (905) 606-1000

Internet: www.rbcinsurance.com

"RBC Insurance is the largest Canadian bank-owned insurance operation and one of the fastest growing in Canada. The company provides a wide range of creditor, life, health, travel, home and auto products and services as well as reinsurance to business clients around the world. RBC Liberty Insurance, its US division, offers traditional and interest-sensitive life insurance products, annuities, health insurance, and related personal financial security products."

Locations

RBC Insurance operates in all Canadian provinces and RBC Liberty Insurance is licensed in 49 states and the District of Columbia.

Careers

As a member company of the RBC Financial Group, which consists of RBC Royal Bank, RBC Centura, RBC Mortgage, RBC Builder Finance, RBC Insurance, RBC Liberty Insurance, RBC Investments, RBC Dain Rauscher, RBC Capital Markets, RBC Global Services, applicants interested in a career with RBC Insurance benefit from the wider range of employment opportunities in the RBC Financial Group. By applying to the RBC Financial Group, they make their profile available to all member companies for recruitment, selection and hire.

The hiring process of RBC Financial Group consists of several steps. It begins with an online application that is designed to match an applicant's "unique skills and interests to possible positions." Potential employees must then pass two interviews.

RBC writes that "your first interview will usually be conducted by an RBC recruiter and may be in person or via phone, depending on your location and availability. RBC Recruitment Services uses behavioral interviewing techniques to assess your suitability for a position. A shortlist of candidates is then presented to the RBC hiring manager for consideration.

"Your second interview will usually be conducted in person by the RBC hiring manager at their office. This will provide you with an ideal opportunity to familiarize yourself with the environment where you may be working.

"The hiring manager, in partnership with the recruiter, will then make the final selection and a conditional offer of employment will be made."

REINSURANCE GROUP OF AMERICA

Headquarters
  1370 Timberlake Manor Parkway
  Chesterfield, Missouri 63017
  Phone: (636) 736-7000
Internet: www.rgare.com

The Reinsurance Group of America is a leader in the global life reinsurance industry. The company provides life reinsurance, risk management, risk assessment, risk transfer, life insurance underwriting and financial services.

Locations

In addition to its United States headquarters, RGA has offices in Buenos Aires, Hong Kong, London, Sydney, Tokyo, Toronto, and affiliated offices in Cape Town, Dublin, Hong Kong, Kuala Lumpur, New Delhi, and Sydney.

Careers

Positions at RGA include actuarial jobs, underwriting jobs, numerous insurance jobs, careers in computer science, actuarial science, mathematics, accounting, law, engineering, business and liberal arts.

ROYAL & SUNALLIANCE

Headquarters
  St. Mark's Court, Chart Way
  Horsham, West Sussex, RH121 1XL
  Phone: (44) (0) (1403) 232 323
Internet: www.royalsunalliance.co.uk

Royal & SunAlliance is one of the world's largest international insurance groups and employs approximately 50,000 individuals in over 50 countries. It is the second largest commercial and personal insurer in the United Kingdom. The three key values of Royal & SunAlliance are truth, trust, and teamwork.

Locations

In the United Kingdom, Royal & SunAlliance has offices in Bournemouth, Bristol, Halifax, Horsham (HQ), Liverpool, London, and Plymouth. The international offices of Royal & SunAlliance are in Argentina, Australia, Brazil, Canada, Chile, China, Colombia, Denmark, Egypt, and the Falkland Islands.

Careers

Royal & SunAlliance operates a virtual university to encourage self-directed learning. The company has "knowledge centers" all over the United Kingdom and supports employees financially and with study leave for the exams which form part of you training scheme.

SSQ FINANCIAL GROUP

Headquarters
2525 Laurier Boulevard
P.O. Box 10500, Station Sainte-Foy
Sainte-Foy, Quebec G1V 4H6
Phone: (418) 683-0554
Internet: www.ssq.ca/en

The SSQ Financial Group is a leading Canadian financial institution with products and services in four sectors of activity: group insurance, investment and retirement, property and casualty insurance, realty management, and promotion and development. The clients of SSQ are also the co-owners of the company.

Location

Quebec City

Careers

"SSQ subsidizes job-related college and university courses for employees. SSQ encourages employees to participate in the workshops offered by different professional associations, notably those provided by the Canadian Institute of Actuaries."

ST PAUL COMPANIES

Headquarters
385 Washington Street
St. Paul, Minnesota 55102-1396
Phone: (651) 310-7911
Internet: www.stpaul.com

The St. Paul Companies provides commercial property-liability insurance and asset management.

Locations

St. Paul is based in Minnesota, with a network of affiliations throughout the United States. "Outside the United States, the company operates through St. Paul International and St. Paul at Lloyd's. St. Paul International provides specialized insurance products and services in the United Kingdom, Ireland, and Canada. Through Global Underwriting, it provides property-liability insurance products for U.S.-based companies with operations outside the United States. St. Paul at Lloyd's underwrites insurance at Lloyd's of London in four principal areas: aviation, marine, global property and specialist personal lines. The St. Paul has discontinued its operations in Argentina, Australia, France, Germany, Mexico, Netherlands, New Zealand, South Africa and Spain."

Careers

"The St. Paul is a company with strong traditions but one that is definitely on the move. The St. Paul is celebrating its 150th anniversary this year. Only 24 companies on the Fortune 500—a mere five percent—have such a long-standing history.... As you explore a job or career, some of the things you will consider are your health and well-being, your financial security, and your work and life balance. At St. Paul, we are proud to offer a comprehensive, high-quality, flexible benefits package that you can personalize to meet your needs now and in the future."

The responsibilities of assistant actuaries or actuaries at St. Paul include leading product rate reviews, new product development, planning and reserving. Candidates for such positions need to have passed "seven Casualty Actuarial Society examinations. Must understand basic ratemaking, loss reserving, and forecasting techniques. Must be familiar with internal and external statistical plans and sources of data. Must be able to program in both mainframe and microcomputer environments as embodied in the actuarial workstation. Must be aware of emerging issues affecting their line of insurance—both technical and product related issues. Must possess good oral and written communication skills. Must be able to deal with people and attain desired results."

STANDARD LIFE

Headquarters
1245 Sherbrooke Street West
Montréal, Quebec H3G 1G3
Phone: (514) 499-8855
Internet: www.standardlife.com

Standard Life has operations in Canada, Ireland, Germany, Austria, Spain, India, and Hong Kong, and has been granted a license to operate in China. The company's products and services include individual and group savings and retirement, group insurance, individual life insurance, and mutual funds.

Locations

In Canada, Standard Life has offices in Calgary and Montreal (HQ).

Careers

"Why work at Standard Life? We care about your satisfaction. Your needs, expectations, and dreams are important to us. By giving you access to the best tools and resources available, we accomplish our mission of helping you grow professionally. We believe that professional development is essential to your success—and that of our organization. Our expression 'employer of choice' is more than just a promise. We have a corporate culture that encourages our employees to achieve excellence."

"Standard Life offers one of the most competitive overall compensation packages on the market: results-based compensation, better benefits than the competition offers, recognition for a job well done, flexible work schedules, continuing development programs, resource center, special events, fitness center."

STATE FARM

Headquarters
One State Farm Plaza
Bloomington, Illinois 61710-0001
Phone: (309) 766-2311
Internet: www.statefarm.com

State Farm is No. 1 in auto and home insurance the United States.

Locations

State Farm has offices throughout the United States and Canada: In Alpharetta (Georgia), Bakersfield (California), Birmingham (Alabama), Bloomington (Illinois) (HQ), Charlottesville (Virginia), Cheshire (Connecticut), Columbia (Montana), Concordville (Pennsylvania), Costa Mesa (California), Dallas (Texas), Duluth (Georgia), Dupont (Washington), Frederick (Maryland), Greeley (Colorado), Jollyville (Texas), Lincoln (Nebraska), Malta (New York), Marshall (Minnesota), Monroe (Louisiana), Murfreesboro (Tennessee), Newark (Ohio), Phoenix (Arizona), Rohnert Park (California), Salem (Oregon), Scarborough (Ontario), Tempe (Arizona), Tulsa (Oklahoma), Wayne (New Jersey), West Lafayette (Indiana), Westlake Village (California), Winter Haven (Florida), and Woodbury (Minnesota).

Careers

The actuarial departments at State Farm develop insurance coverages and rates in P/C, life, and heath.

An actuarial technician in the P/C department would deal with private and commercial auto insurance pricing, and recommend and implement pricing which satisfies company financial goals. "Actuaries work as a team to research and develop new products and to estimate future premiums, losses, and expense costs. Actuaries also gather and analyze financial and statistical data, assure compliance with insurance regulations and statutes, and represent State Farm at industry meetings and on actuarial committees."

An Actuarial technicians in life and health will "research and develop new products and conduct investment analysis and modeling for the financial position of the company." They also "gather and analyze financial and statistical data, assure compliance with insurance regulations and statutes, and represent State Farm at industry meetings and on actuarial committees."

The company also encourages its actuarial trainees to become Fellows in the Casualty Actuarial Society with the assistance of its competitive exam support program.

Actuarial technicians must have passed at least one actuarial exam and a Bachelor's degree with a high overall GPA [grade point average] in actuarial science, mathematics, or statistics is strongly desired. Actuaries at State Farm are expected to have strong analytical, problem-solving skills, and communication skills. They must have a "strong desire and commitment to pursue the actuarial exams towards the attainment of the FCAS designation."

SUN LIFE FINANCIAL

Headquarters
  150 King Street West
  Toronto, Ontario
  Canada M5H 1J9
  Phone: (416) 979-9966
Internet: www.sunlife.com

Sun Life Financial is a leading financial services organization with operations in key markets around the world.

  Sun Life Financial offers a diverse range of financial products and services in wealth management and protection. "Wealth management includes asset management, mutual funds, pension plans and products, and annuities operations. Protection includes whole life, term life, universal life, unit-linked life and corporate-owned life insurance for individuals. As well, life, health and disability insurance products are offered to group clients." Worldwide, Sun Life Financial has approximately 15,000 employees.

Locations

Bermuda, Canada, Chile, China, Hong Kong, India, Indonesia, Ireland, Philippines, United States, United Kingdom.

Careers

"Sun Life Financial is a leading international financial services organization providing a diverse range of wealth accumulation and protection products and services to individuals and corporate customers.

  "A company is only as good as its people. At Sun Life Financial, one of our primary core values dictates that we pursue Excellence through the people we employ and the work that they do. As a world-class financial services organization, we recognize that the contributions made by our employees are vital to our success. We are constantly seeking high-caliber individuals who will bring excellence, talent and a special energy to our dynamic, growing family of operations.

  "We offer a diverse range of exciting career opportunities, supported by extensive training and development programs to help our employees reach their full potential."

SWISS REINSURANCE

Headquarters
175 King St.
Armonk, New York 10504
Phone: (877) 794-7773
Internet: www.swissre.com

Swiss Reinsurance America Corporation is a division of Swiss Re, a worldwide reinsurance company with offices in Africa, Asia, Australasia, Europe, and North and South America.

Locations

Armonk (New York) and branch offices in Atlanta, Boston, Philadelphia, and San Francisco.

Careers

Swiss Re sees reinsurance as an evolving industry and employs "innovative, forward-looking people who know the insurance industry's demands." The company creates a flexible environment that promotes creativity. Swiss Re "aims to be the pioneer that shapes reinsurance to reflect client requirements."

A senior actuary at Swiss Re works "as part of a multi-disciplined deal team developing pricing solutions. In addition to being involved in the analysis of industry-specific data." Part of the actuary's responsibility is to "obtain and interpret industry loss data and perform actuarial modeling of the risk process underlying the deals," as well as developing rating tools and undertaking research in new areas of operation. A senior actuary is expected to have a degree in mathematics or actuarial science, at least three years experience in insurance or reinsurance pricing, be personable, and have good communication skills.

Swiss Re also employs marketing actuaries. In its Latin American division, for example, a marketing actuary in the life and health "will add value for clients through a market-oriented approach to knowledge and risk transfer." Responsibilities include carrying out "mortality and morbidity studies in Latin American markets," coordinating "with internal units to gather information on products, pricing and best-practice guidelines," developing "innovative Life & Health products for Latin America which meet different needs based on the age, economic and social profile of individual and group consumers." Fluency in Spanish, English and German are required. So are "experience in actuarial mathematics and statistics, experience in creating statistical models for business needs, the ability to present technical information in a clear, concise and confident manner, innovative problem-solving skills, a genuine interest in market needs, ability to perform in multicultural teams, and high commitment."

TRANSATLANTIC HOLDINGS

Headquarters
New York/Corporate Office
80 Pine Street
New York, New York 10005
Phone: (212) 770-2000
Internet: www.transre.com

Transatlantic Holdings, Inc. is a leading international reinsurance organization. Its subsidiaries are Transatlantic Reinsurance Company, Putnam Reinsurance Company, and Trans Re Zurich. These companies offer a full range of property and casualty reinsurance products, with an emphasis on specialty risks.

Locations

Transatlantic has offices in Buenos Aires, Chicago, Hong Kong, Johannesburg, London, Miami (serving Latin America and the Caribbean), New York (HQ), Paris, Rio de Janeiro, Sydney, Shanghai, Toronto, Warsaw, Tokyo, and Zurich.

Careers

Depending on the position, candidates for employment at Transatlantic need a property and casualty insurance background. They are also expected to have strong communication and interpersonal skills. Required actuarial skills and experience include pricing using catastrophe models.

WILLIS GROUP HOLDINGS

Headquarters
Willis Group Holdings Limited
7 Hanover Square
New York, New York 10004-2594
Phone: (212) 344-8888
Internet: www.willis.com

Willis Group Holdings is a leading global insurance broker, developing and delivering professional insurance, reinsurance, risk management, financial and human resource consulting and actuarial services to corporations, public entities and institutions around the world. Willis has particular expertise in serving the needs of clients in such major industries as construction, aerospace, marine and energy.

Locations

Willis has over 300 offices in more than 100 countries and its global team of 13,000 associates serve clients in 180 countries.

Careers

Willis seeks individuals who are innovative thinkers, possess a high degree of integrity, subscribe to a knowledge-sharing philosophy, value collaboration and teamwork, pursue continuous learning and personal development, are performance-achievers, entrepreneurial in spirit, and take pride in their organization.

RECIPROCITY

The Faculty and Institute have signed mutual recognition agreements with several actuarial organizations: the Institute of Actuaries of Australia, the Canadian Institute of Actuaries, the Society of Actuaries of the United States, the Institute of Actuaries of Japan, and the *Groupe Consultatif* of the European Union.

The agreement describes the process, country by country, by which actuaries in the countries involved can become members of the actuarial societies in the other participating countries. The document, summarized in this appendix, is the official description of the reciprocity agreement and should be consulted for specific details.

The agreement says, in essence, that actuaries who have become Fellows of a national actuarial society by the normal route (having passed the necessary examinations), and who are members in good standing (having paid the annual membership fee in their home country), meet the professionalism requirements of the guest country, fulfill the necessary residency requirements, and intend to practice in the guest country, can do so by reciprocity. Anyone interested in taking advantage of this agreement should consult the relevant documents published on the Faculty of Actuaries and Institute of Actuaries and *Groupe Consultatif* websites:

1. www.actuaries.org.uk/files/pdf/worldwide/mutual_recog.pdf
2. www.gcactuaries.org/documents/recognition.pdf

England, Scotland and Australia

Fellows of the Institute of Actuaries of Australia in good standing can become Fellows of the Faculty of Actuaries and the Institute of Actuaries on the following conditions:

▶ They have become full membership of the IAAust by examination and not in recognition of membership of another actuarial association.

▶ They wish to become practicing actuaries in the United Kingdom or Republic of Ireland or intend to advise on UK or Irish business.

▶ They have at least three years' recent appropriate practical experience of which at least one year was of UK or the Republic of Ireland business.

▶ They have attended an approved professionalism course.

Conversely, the Institute of Actuaries of Australia will admit to Accredited Member status of the IAAust Fellows of the Faculty of Actuaries or the Institute of Actuaries who wish to pursue actively the profession of actuary in Australia, provided that they satisfy the following conditions:

▶ They have qualified as Fellows of the Faculty of Actuaries or the Institute of Actuaries through examination.

▶ They have been resident in Australia for at least 6 months.

▶ They gained suitable experience in local actuarial practice.

▶ They have passed a recognized professionalism course within the previous 5 years or earlier at the discretion of the Committee, or any other course approved by the Committee.

Applicants who meet these conditions will automatically be recommended to Council of the Institute of Actuaries of Australia for membership.

England, Scotland and Canada

Canadian actuaries intending to practice in the United Kingdom or Republic of Ireland can become Fellows of the Faculty of Actuaries and the Institute of Actuaries on conditions similar to those for Australian actuaries, provided that they are full members of the Canadian Institute of Actuaries through examination from the Society of Actuaries, the Casualty Actuarial Society or the Institute of Actuaries of Australia.

The Canadian Institute of Actuaries will, in turn, admit as a permanent Member of the Canadian Institute of Actuaries, a Fellow of the Faculty of Actuaries or the Institute of Actuaries, who wishes to pursue actively the profession of actuary in Canada, on conditions similar to the Australian case. Applicants must also have passed the Practice Education Course (PEC) of the Canadian Institute of Actuaries. As stipulated in the reciprocity agreement, this course "may be attended following 12 months of relevant Canadian experience; must have completed at least 12 hours of Canadian Institute of Actuaries-approved professional development (PD) in the 12 months prior to the application for Member status. They are required to demonstrate that they have completed a three-year period of practical actuarial work experience, including at least 18 months of specifically

Canadian practical actuarial work experience within the three-year period immediately prior to their application for Member status. They must disclose to the Canadian Institute of Actuaries any public disciplinary sanctions that have been imposed against them by any actuarial organization of which they are a Fellow (or equivalent). This information is taken into account when determining whether the applicants should be granted Membership status in the Canadian Institute of Actuaries.

England, Scotland and the United States

The agreement between the Faculty of Actuaries and the Institute of Actuaries and the Society of Actuaries is similar to the two previous examples. To become an accredited member of the Society of Actuaires, an applicant must fulfill the following conditions:

► Have attained full membership of the Faculty of Actuaries or the Institute of Actuaries by examination and not in recognition of membership of another actuarial association.

► Be a Fellow in good standing of the Canadian Institute of Actuaries, or Member in good standing of the American Academy of Actuaries, or full member in good standing of other actuarial associations designated from time to time by the Society of Actuaries Board of Governors.

► Have attended and passed the Society of Actuaries Fellowship Admissions Course, or its equivalent as recognized by the Society of Actuaries, in the five years prior to application.

► Have satisfied the Society of Actuaries Professional Development requirements, or its equivalent as recognized by the Society of Actuaries, in the five years prior to application.

England, Scotland and Japan

In the case of Japan, reciprocity of more limited. The agreement states that "the Institute of Actuaries will to admit to its *Affiliate* status any Fellow of the Institute of Actuaries of Japan (IAJ) who submits an application form to the Faculty of Actuaries or the Institute of Actuaries and pays the required fee, subject to any conditions prescribed for such status. The IAJ will admit to its "Kenkyu-Kaiin" status any Fellow of the Faculty of Actuaries or Institute of Actuaries who submits an application form to the IAJ and pays the required fee, subject to any conditions prescribed for such status."

Affiliates of the Faculty of Actuaries and Institute of Actuaries have no voting rights, and it is agreed as part of this agreement that Fellows of the Faculty of

Actuaries or the Institute of Actuaries who become Kenkyu-Kaiin will have no voting rights in the IAJ.

Fellows of either organization taking advantage of the stated program of the other organization "may attend all meetings and programs of the other organization at the same rate charged to members, although they may be excluded from business meetings at which membership votes are to be taken."

Fellows of the Faculty of Actuaries or the Institute of Actuaries who become Kenkyu-Kaiin become non-voting members of the IAJ and are subject to professional requirements of the IAJ.

England, Scotland and the European Union

The Faculty and Institute are signatories to the *Groupe Consultatif* Agreement on the Mutual Recognition of Qualifications. Under this agreement there are two routes by which a full member of one European actuarial association can become a Fellow of another European actuarial association. (1) Undergo an supervised adaptation period so that the applicant has at least three years' practical experience in total, of which at least one year is in the host country. (2) Pass an aptitude test. The applicant has the choice of routes.

"An applicant for Fellowship of the Faculty of Actuaries or the Institute of Actuaries who to undertake an adaptation period must be under the supervision of a Fellow of the Faculty of Actuaries or the Institute of Actuaries. The supervisor should have worked as an actuary in the United Kingdom for at least three out of the past five years and have completed a program of Continuing Professional Development in accordance with the Faculty and Institute scheme. Applicants must send their application to the Faculty of Actuaries and Institute of Actuaries which will administer the process. Fellows of the Faculty of Actuaries or Institute of Actuaries working in another Member State of the European Union or Switzerland which has an actuarial association which is a member of the *Groupe Consultatif* are required to join their host association. They should contact the host association for details on the process for achieving this."

The European Union

The *Groupe Consultatif Actuariel Europeen* has established reciprocity agreements for the recognition of actuarial qualification between the following national associations of actuaries in Europe. These include the following Deutsche Aktuarvereinigung (Germany), Aktuarvereinigung Österreichs (Austria), Association Royale des Actuaires Belges / Koninklijke Vereniging, van Belgische Aktuarissen (Belgium), Den Danske Aktuarforening (Denmark), Instituto de Actuarios Españoles (Spain), Collegi d'Actuaris de Catalunya (Spain), Suomen Aktuaariyhdistys (Finland), Association des Actuaires Diplômés de l'Institut de Science

Financière et d'Assurances (France), Institut des Actuaires Français (France), Union Strasbourgeoise des Actuaires (France), Association of Greek Actuaries (Greece), Society of Actuaries in Ireland, Istituto Italiano degli Attuari (Italy), Association Luxembourgeoise des Actuaires (Luxemburg), Het Actuarieel Genootschap (Netherlands), Instituto dos Actuarios Portugueses (Portugal), The Faculty of Actuaries (Scotland), Institute of Actuaries (England), Svenska Aktuarieföreningen (Sweden), as well as Den Norske Aktuarforening (Norway), and Felag Islenskra Tryggingast Aerdfraedinga (Iceland).

ACTUARIAL WEBSITES

North American Organizations

1. Actuarial Foundation
   Internet: www.actuarialfoundation.org
2. American Academy of Actuaries
   Internet: www.actuary.org
3. American Society of Pension Actuaries
   Internet: www.aspa.org
4. Canadian Institute of Actuaries
   Internet: www.actuaries.ca
5. Casualty Actuarial Society
   Internet: www.casact.org
6. Conference of Consulting Actuaries
   Internet: www.ccactuaries.com
7. Society of Actuaries
   Internet: www.soa.org

Other National Organizations

1. Aktuarvereinigung Österreichs (Austria)
   Internet: www.avoe.at
2. Asociacion Mexicana de Actuarios (Mexico)
   Internet: www.amac.org.mx
3. Association Suisse des Actuaires (Switzerland)
   Internet: www.actuaries.ch

4. Actuarial Society of Hong Kong
   Internet: www.actuaries.org.hk
5. Actuarial Society of India
   Internet: www.actuariesindia.org
6. Actuarial Society of Malaysia
   Internet: www.actuaries.org.my
7. Actuarial Society of South Africa
   Internet: www.assa.org.za
8. Association Royale des Actuaires Belges (Belgium)
   Internet: www.actuaweb.be
9. Australian Actuarial Society
   Internet: www.acted.com.au
10. Den Danske Aktuarforening (Denmark)
    Internet: www.aktuarforeningen.dk
11. Deutsche Aktuarvereinigung (Germany)
    Internet: www.aktuar.de
12. Den Norske Aktuarforening (Norway)
    Internet: www.aktfor.no
13. Faculty and Institute of Actuaries (UK)
    Internet: www.actuaries.org.uk
14. Fédération Française des Actuaires (France)
    Internet: www.actuaires.com.fr
15. Groupe Consultatif Actuariel Europeen (European Union)
    Internet: www.gcactuaries.org
16. Het Actuarieel Genootschap (Netherlands)
    Internet: www.ag-ai.nl
17. Institut des Actuaires Français (France)
    Internet: www.institutdesactuaires.com
18. Institute of Actuaries of Japan
    Internet: www.iaj-web.or.jp
19. Instituto Brasileiro de Atuária (Brazil)
    Internet: www.actuary-iba.org.br
20. Instituto de Actuarios Españoles (Spain)
    Internet: www.actuarios.org
21. International Actuarial Association
    Internet: www.actuaries.org
22. International Association of Consulting Actuaries
    Internet: www.iacactuaries.org
23. Israel Association of Actuaries
    Internet: www.actuaries.org.il

24. Istituto Italiano degli Attuari
    Internet: www.italian-actuaries.org
25. Japanese Society of Certified Pension Actuaries
    Internet: www.jscpa.or.jp
26. New Zealand Society of Actuaries
    Internet: www.actuaries.org.nz
27. Portuguese Institute of Actuaries
    Internet: www.actuarial.pt
28. Society of Actuaries in Ireland
    Internet: www.actuaries-soc.ie
29. Suomen Aktuaariyhdistys (Finland)
    Internet: www.actuary.fi
30. Svenska Aktuarieföreningen (Sweden)
    Internet: www.aktuarieforeningen.com
31. The Chinese Actuarial Club (China)
    Internet: www.chinese-actuary.org

Recruiting Agencies

1. Acsys Inc. (West Des Moines, Iowa)
   Internet: www.acsysinc.com
2. ACTEX Actuarial Recruiting (Winsted, Connecticut)
   Internet: www.actexmadriver.com
3. Actuarial Careers Inc. (White Plains, New York)
   Internet: www.actuarialcareers.com
4. Actuarial Search Associates (Venice, California)
   Internet: www.actuarialsearch.com
5. Acumen Resources (London, UK)
   Internet: www.acumen-resources.com
6. Andover Research Ltd. (New York, New York)
   Internet: www.andoverresearch.com
7. CPS Inc. (Boston, Massachusetts)
   Internet: www.cps4jobs.com
8. D. W. Simpson and Company (Chicago, Illinois)
   Internet: www.dwsimpson.com
9. Inside Careers Guide (UK)
   Internet: www.insidecareers.co.uk
10. Jacobson Associates (Chicago, Illinois)
    Internet: www.learn2.com
11. Mid America Search (West Des Moines, Iowa)
    Internet: www.midamericasearch.com

12. Pinnacle Group (Portsmouth, New Hampshire)
    Internet: www.pinnaclejobs.com
13. Pryor Associates (Hicksville, New York)
    Internet: www.ppryor.com
14. SC International Ltd. (Downers Grove, Illinois)
    Internet: www.scinternational.com
15. Stewart Search Advisors LLC (Portsmouth, New Hampshire)
    Internet: www.StewartSearch.com and **www.ActuarialFutures.com**

ACTUARIAL SYMBOLS

True to the origin of their name, actuaries use an extensive list of special symbols for their work. It's a kind of cleverly devised shorthand for actuarial objects and functions. The notation is based on principles for construction adopted by the Second International Congress of Actuaries in London in 1898. The list is modified and updated from time to time with the approval of the Permanent Committee of Actuarial Notations of the International Actuarial Association. Appendices 3 and 4 of the standard reference on actuarial mathematics by Bowers et al. (*see* Reference 5, Appendix F) contains a full list of symbols. The following list gives an indication of the kind of symbols involved. Many of these symbols occur in the sample Questions and Answers in Chapter 2. For missing symbols used in these examples, we refer to Reference 5, Appendix F for their definition and explanation. Relevant financial symbols are also discussed in Section 2.

- ► The symbol $a_{\overline{n}|}$ denotes the value of an annuity of one dollar per year for n years, payable at the end of each year.
- ► The symbol $\ddot{a}_{\overline{n}|}$ denotes of value of an annuity of one dollar per year for n years, payable at the beginning of each year.
- ► The symbol $a_{\overline{n}|i}$ denotes the value of an annuity of one dollar per year for n years at i percent interest per year, payable at the end of each year.
- ► The symbol $\ddot{a}_{\overline{n}|i}$ denotes the value of an annuity of one dollar per year for n years at i percent interest per year, payable at the beginning of each year.
- ► The symbol $\overline{a}$ denotes an annuity payable continuously at the rate of one dollar per year.
- ► The symbol $\ddot{a}_{xy}$ denotes an annuity payable during the joint lives of (x) and (y), payable at the beginning of each year.

▶ The symbol $\ddot{a}_{\overline{xy}}$ denotes an annuity payable as long as one of (x) and (y) is alive, payable at the beginning of each year.

▶ The symbol $\ddot{a}_{x:\overline{n}|}$ denotes an annuity payable during the joint duration of the life of (x) and a term of n years.

▶ The symbol $_nd_x$ denotes the expected number of deaths between the ages x and $x+n$.

▶ The symbol $\overset{\circ}{e}_x$ denotes the average remaining lifetime at age x.

▶ The symbol l_x denotes the expected number of survivors to age x from l_0 newborns.

▶ The symbol p_x denotes the probability that life (x) will reach age $x+1$.

▶ The symbol $_tp_x$ denotes the probability that life (x) will survive the next t years.

▶ The symbol P_x denotes the annual premium of a whole life policy of one dollar, payable upon the death of x, with the first premium payable when the policy is issued.

▶ The symbol P_{xy} denotes the annual premium of a whole life policy of one dollar, payable upon the death of x, with the first premium payable when the policy is issued.

▶ The symbol $P_{\overline{xy}}$ denotes the annual premium of a whole life policy of one dollar, payable upon the first death of x or y, with the first premium payable when the policy is issued.

▶ The symbol q_x denotes the probability that life (x) will die within the next year.

▶ The symbol $_tq_x$ denotes the probability that life (x) will die within t years.

▶ The symbol $s(x)$ denotes the probability that a newborn will reach age x.

▶ The symbol (x) denotes a living person age x.

▶ The symbols (xy) denotes two living persons age x and y, respectively.

BIBLIOGRAPHY

[1] *Actuarial Career Planner*. The Society of Actuaries, Schaumburg, Illinois, 1998.

[2] Alexander, D., *Steps Forward in China,* International Section Newsletter, Number 24, February 2001, The Society of Actuaries, Schaumburg, Illinois.

[3] *Basic Education Catalog* (Fall 2002). The Society of Actuaries, Schaumburg, Illinois, 2002.

[4] *Basic Education Catalog* (Spring 2003). The Society of Actuaries, Schaumburg, Illinois, 2003.

[5] Bowers, N. L., Gerber, H. U., Hickman, J. C., Jones, D. A., and Nesbitt, C. J., *Actuarial Mathematics* (2nd Edition). The Society of Actuaries, Schaumburg, Illinois, 1997.

[6] Brown, R. L., *The Globalisation of Actuarial Education*, British Actuarial Journal, Volume 8, Number 1, 2002, Pages 1–3.

[7] *CAS Survey on Professional Skills*, The Casualty Actuarial Society, Arlington, Virginia, 2002.

[8] *CAS Syllabus*. The Casualty Actuarial Society, Arlington, Virginia, 2002.

[9] *Encyclopaedia Britannica* (11th Edition), Volume XXII, London, 1911.

[10] Jones, B. L., *An Introduction to Actuarial Models and Modeling (An Interactive Approach)*. Actex Publications, Winsted, Connecticut, 2000.

[11] Kellison, S. G., *The Theory of Interest* (2nd Edition). Irwin McGraw-Hill, Boston, 1991.

[12] Klugman, S. A., Panjer, H. H., and Willmot, G. E., *Loss Models: From Data to Decisions*. John Wiley and Sons, New York, 1998.

[13] Krantz, L., and Lee, T., *Jobs Rated Almanac, 2002* (6th Edition). Barricade, 2002.

[14] *Learn₂*, E-Learning Online Superstore, Internet: www.jacobson-associates. com, 2003.

[15] Morgan, E., *Love it or hate it*, The Actuary, Staple Inn Actuarial Society, July, 2001.

[16] Narvell, J. C., *India in Transition*, Actuarial Review (May 2003), The Casualty Actuarial Society, Arlington, Virginia, 2002.

[17] Ott, R. L., and Longnecker, M., *Statistical Methods and Data Analysis* (5th Edition). Duxbury, Pacific Grove, California, 2001.

[18] Perryman, F. S., *International Actuarial Notation*, Proceedings of the Casualty Actuarial Society, Volume 36, Number 66, 1949, Pages 123-131.

[19] *Principles Underlying Actuarial Science.* CAS Committee on Principles and SOA Committee on Actuarial Principles. Exposure Draft, October 15, 1999.

[20] Szabo, F. E., *Linear Algebra: An Introduction using Maple*. Harcourt/ Academic Press, Boston, 2002.

[21] Wackerly, D. D., Medenhall, W., and Scheaffer, R. L, *Mathematical Statistics with Applications* (5th Edition).Wadsworth Publishing Company, Belmont, California, 1996.

[22] Warren, W. S., *The Physical Basis of Chemistry* (2nd Edition). Harcourt/ Academic Press, Boston, 2000.

[23] Zima, P., and Brown, R. L., *Mathematics of Finance* (5th Edition). McGraw-Hill Ryseron, Toronto, 2001.

INDEX